Mobile Robotic Car Design

Pushkin Kachroo

Patricia Mellodge

D1160952

McGraw-Hill

New York   Chicago   San Francisco   Lisbon   London   Madrid
Mexico City   Milan   New Delhi   San Juan   Seoul
Singapore   Sydney   Toronto

Cataloging-in-Publication Data is on file with the Library of Congress

1 2 3 4 5 6 7 8 9 0 DOC/DOC 0 1 0 9 8 7 6 5 4

ISBN 0-07-143870-X

The sponsoring editor for this book was Judy Bass and the production supervisor was Pamela A. Pelton. It was set in Century Schoolbook by MacAllister Publishing Services, LLC. The art director for the cover was Anthony Landi.

Printed and bound by RR Donnelley.

McGraw-Hill books are available at special quantity discounts to use as premiums and sales promotions, or for use in corporate training programs. For more information, please write to the Director of Special Sales, McGraw-Hill Professional, Two Penn Plaza, New York, NY 10121-2298. Or contact your local bookstore.

PIC, MPLAB, PRO MATE, and PICSTART are registered trademarks of Microchip Technology Inc. MATLAB and Simulink are registered trademarks of The MathWorks, Inc. Code Composer Studio is a registered trademark of Texas Instruments Incorporated.

This book is printed on recycled, acid-free paper containing a minimum of 50 percent recycled, de-inked fiber.

This book is dedicated to my wife Anjala Krishen,
who has always supported me in every way.   P.K.

About the Authors

Pushkin Kachroo, Ph.D., P.E., is a noted robotics theorist and developer. The author of three books and more than 75 papers published in technical journals and conferences, he is an Associate Professor in the Bradley Department of Electrical and Computing Engineering at Virginia Polytechnic Institute, Blacksburg, Virginia. He earned his doctorate at the University of California, Berkeley; his master's degree at Rice University; and his bachelor's degree at the Indian Institute of Technology, Bombay. He obtained his Professional Engineering license from the state of Ohio in 1995.

Patricia Mellodge specializes in mobile robot control. An electrical engineering graduate of the University of Rhode Island, she designed analog control circuits for Optigain, Inc., before joining the Bradley Department of Electrical and Computing Engineering at Virginia Polytechnic Institute. In 2002, she received her master's degree and she is currently pursuing her Ph.D.

Contents

Preface

We envisage two main types of readers for this book: a hobbyist who wants to build a robot, and a university-level student or researcher who wants to build a robotic experimental vehicle and conduct research experiments on it. This book describes the complete details of how to build a robotic car. This information should be clear enough for a hobbyist to be able to build a robot from scratch. We have provided detailed photographs, schematics, and a list of vendors for parts. We provide a stepwise procedure for the experimenter or hobbyist to build the robotic car. We have also provided the software code for the processor and the *digital signal processor* (DSP) and have given an explanation of what the code does and why. In addition, we have provided general information on various components of the robotic car. For instance, we have a chapter on DC motors and servos that starts by describing the physics behind the electromechanical motion, and follows by describing how to control those devices. Finally, we give mathematical models for the same devices. A hobbyist can choose to skip the mathematical parts of the book and concentrate on the construction of the robot, as well as the control algorithms that can be programmed into the robot, without worrying too much about their derivations.

The following are some books that can be used as general references for various topics covered in the book. We have chosen these books so the reader of this book can pick up the needed concepts as easily as possible.

Topic	Reference Book Information
Electric circuits	John O'Malley
	Schaum's Outline of Basic Circuit Analysis
	McGraw-Hill Trade, 2nd edition, 1992
Electronics	Allan R. Hambley
	Electronics
	Prentice Hall, 2nd edition, 1999
Nonlinear control	Jean-Jacques Slotine and Weiping Li
	Applied Nonlinear Control
	Prentice Hall, 1990
Modeling and control	Charles L. Phillips and Royce D. Harbor
	Feedback Control Systems
	Prentice Hall, 4th edition, 1999

In this book, we provide many more references for the various topics we discuss. However, the books mentioned previously can help the reader review or learn some basic material that is a prerequisite for the reader to comprehend different sections. However, as we mentioned before, even without having a background in robotics, readers can learn various things from different sections. For example, a hobbyist may build a car without worrying about the mathematical sections of the book.

This robotic car was the brainchild of the first author, Dr. Pushkin Kachroo, who has worked on this project for many years. He worked on vehicle traction control during his Ph.D. work at the University of California at Berkeley under Dr. Masayoshi Tomizuka. At Virginia Tech, he started building the small-scale car for experimentation in the context of intelligent vehicles. He obtained funding on this project with the help of Ray Pethtel at the Virginia Tech Transportation Institute, and later with Ray Pethtel and William "Bill" Green in the department of Architecture at Virginia Tech. Dr. Kachroo has had many students who worked on this project and did their graduate research with him on this topic. Some of them are

- Nikolai Schlegel, *Autonomous Vehicle Control Using Image Processing*, M.S. thesis, Virginia Tech, 1998, http://scholar.lib.vt.edu/theses/available/etd283421290973280/unrestricted/etd.pdf
- Richard D. Henry, *Automatic Ultrasonic Headway Control for a Scaled Robotic Car*, M.S. thesis, Virginia Tech, 2001, http://scholar.lib.vt.edu/theses/available/etd-12182001-232108/unrestricted/thesis.pdf
- Patricia Mellodge, *Feedback Control for a Path Following Robotic Car*, M.S. thesis, Virginia Tech, 2002, http://scholar.lib.vt.edu/theses/available/etd-05022002-143530/unrestricted/etd.pdf
- Eric N. Moret, *Dynamic Modeling and Control of a Car-Like Robot*, M.S. thesis, Virginia Tech, 2003, http://scholar.lib.vt.edu/theses/available/etd-03242003-120642/unrestricted/thesis_ENMoret.pdf

Moreover, Mark Morton is also finishing his M.S. thesis on traction control experiments on the scaled robotic car, and Chris Corey Howells is finishing his M.S. thesis on designing collision-avoidance controllers using game theory. The second author, Patricia Mellodge, is working on her Ph.D using these robotic cars for her experiments.

We hope that the readers of this book, whether they are hobbyists or researchers, have as much fun reading, building, studying, and researching as we have with these robotic cars.

Introduction

This book is intended to teach you about mobile robotic cars. By the time you finish this book, you will understand enough about them to be able to design and build your own. The question is, why would you want to do such a thing?

Motivation

The theory and design of mobile robotic cars are the subjects of much research in universities and companies around the world. One major application is the design of cars and trucks that drive themselves. Why automate the driving task? One of the major reasons is safety. According to the U.S. Department of Transportation, in 2000 there were approximately 6,394,000 police-reported motor vehicle traffic crashes, resulting in 3,189,000 injuries and 41,821 lives lost (U.S. DOT 2001). Accidents on our roadways not only cause injuries and fatalities, but also have a huge economic impact (Reed 1992). Many accidents are caused by human error. Eliminating this error can reduce the number of injuries and fatalities on our roadways.

Human driving error may be caused by a number of factors, including fatigue and distraction. During long drives on the highway, the driver must constantly monitor the road conditions and react to them over an extended period of time. Such constant attentiveness is tiring and the resulting fatigue may reduce the driver's reaction time. Additionally, the driver may be distracted from the task of driving by talking with other passengers, tuning the radio, or using a cell phone. Such distractions may also lead to accidents. The U.S. Department of Transportation has released statistics indicating that driver distraction was a factor in 11 percent of fatal crashes, and in 25 to 30 percent of injury and property-damage-only crashes in 1999 (Utter 2001). On the other hand, a car capable of driving itself can allow the occupants to perform nondriving tasks safely while traveling to their destination.

Another reason to automate cars is to alleviate congestion on the highways. A method called "platooning" allows cars to drive at highway speeds while only a few feet apart. Because the electronics on the car can respond faster than a human can, cars would be able to drive much closer together. This would result in a safer, more efficient use of the existing highways.

How do automatic cars relate to robotic cars? The answer is that robotic cars are used for scale-model testing. Scale-model testing allows for the safe and inexpensive implementation of prototype designs. It is more cost effective and safer to use a small-scale robotic car rather than a full-scale car for initial testing. Testing (and repairing after the inevitable crashes!) is easier on a small-scale vehicle. A full-scale prototype requires a full-scale roadway on which to test, rather than the relatively small area needed for scale-model testing. The robotic car described in this book was built for this use as part of the *Flexible Low-cost Automated Scaled Highway* (FLASH) Project at the Virginia Tech Transportation Institute.

Other applications for mobile robotic cars involve sending them where humans cannot or should not go. NASA has programs that develop robots to navigate other planets. It is safer and less expensive to send a relatively small, mobile robot to the harsh environment of Mars than to develop transportation and living quarters for several astronauts. Similarly, mobile robots can enter burning buildings to locate someone or something. They can navigate battlefields to search for mines. They can seek out and deactivate bombs. The usefulness of these little machines is limited only by your imagination.

Although all of these applications are important and useful, there is another (and maybe more important) reason to design and build mobile robots: It's fun! For the hobbyist, robots are an endless source of ideas for tinkering. Designing and building a robot allows you to be involved in many different interesting areas: mechanical design, electronic design, computer programming, mathematical modeling, and control design.

In terms of mechanical design, a robot is made up of several different parts that fit together. These parts must be designed to ensure that the robot moves correctly and efficiently. For example, the motor driving the robot must be geared so the wheels turn at the desired speed and with the desired torque. The wheels for steering must be attached to a motor so they turn to the desired angles. Sensors must be located so they can collect data from the robot or its environment. All of the circuitry must be packaged and placed so it doesn't interfere with the sensors or the robot's movement. The mechanical design involves making sure all of the various parts fit together correctly and move efficiently.

Mobile robots are loaded with electronic circuitry. Circuits are needed to actuate the motors and operate the sensors. They are also needed to act as an interface between the computer (the "brain") and the rest of the robot. For example, in order to drive the motor, the computer generates a *pulse width mod-*

ulated (PWM) signal. But the computer's output is not compatible with the motor, so an interface circuit known as an H-bridge is needed between the two components.

The brain of a robot is its computer. The computer must be programmed so that the robot knows what to do. If the robot is to follow a line, it must know where the line is and how to steer itself accordingly. The computer is programmed with algorithms that use the sensor data to determine how the robot must move.

For the mathematically minded, the modeling of the robot provides great opportunities. A model is a mathematical description of the robot's behavior. To make a robot move in a certain manner, you must know how it will respond when it's told to move. For example, when the computer gives a command to turn the wheels to 15 degrees and drive at 3 feet per second, the model predicts the robot's movement. The model is necessary when you design a control algorithm to make the robot perform a desired task.

This leads us to the control design. The purpose of the control algorithm is to tell the robot how to move based upon the data it receives from the sensors. The controller can be likened to a person driving a car. When you drive, you look at the road, the cars around you, and pedestrians. You also listen for sirens, honking horns, and other noises. This is the data that you are collecting from your sensors, which, in this case, are your eyes and ears. Based on this data, you provide the necessary input to make the car go where you want; for example, you step on the throttle or the brake and you turn the steering wheel. The process that you go through, probably without thinking about it, is a control algorithm.

As you can see, designing mobile robots entails many different aspects. This book covers all these areas so that you can decide for yourself on which area to concentrate.

Autonomous Vehicles

The inventions of the *integrated circuit* (IC) and, later, the microcomputer were major factors in the development of the electronic control of vehicles. The importance of the microcomputer cannot be overemphasized as it is the "brain" that controls many systems today. For example, in a car's cruise control system, the driver sets the desired speed and enables the system by pushing a button. A microcomputer then monitors the actual speed of the vehicle using data from velocity sensors. The actual speed is compared to the desired speed, and the controller adjusts the throttle as necessary.

A completely autonomous vehicle is one in which a computer performs all the tasks that the human normally would. Ultimately, this would mean entering the task into a computer and releasing the system. At that point, the robotic car would take over and perform the task with no human input. The robot would be able to sense its environment and determine its movement accordingly.

In the case of standard automobiles, this scenario would require many different technologies: lane detection to aid in passing slower vehicles or exiting a highway; obstacle detection to locate other cars, pedestrians, and animals; adaptive cruise control to maintain a safe speed; collision avoidance to avoid hitting obstacles in the roadway; and lateral control to maintain the car's position on the roadway. In addition, sensors would be needed to alert the car to road or weather conditions to ensure safe traveling speeds. For example, the car would need to slow down in snowy or icy conditions.

Completely automating a vehicle is a challenging task. However, advances have been made in the individual systems. For example, cruise control is common in cars today. Adaptive cruise control, in which the car slows if it detects a slower moving vehicle in front of it, is available on higher-end models. In addition, some cars come equipped with sensors to determine if an obstacle is near and will sound an audible warning to the driver when it is too close. As the technology advances, robots will become capable of performing more tasks without human intervention.

Organization of This Book

This book is organized into two parts. Part I, "Hardware Implementation," consists of Chapters 2 through 7 and provides a description of the hardware implementation of a mobile robotic car. It gives details on each subsystem of the car and how to build one. The chapters of Part I are as follows:

- Chapter 2 describes the overall system structure and the components used.
- Chapter 3 shows, step by step, the construction of the car.
- Chapter 4 gives details of the sensors and the circuitry used to interface with them.
- Chapter 5 describes the car's actuation: the motor and servo.
- Chapter 6 provides programming details for the low-level control of the car.
- Chapter 7 describes the programming of the car's main processor.

Part II, "Theory of Mobile Robots," consists of Chapters 8 through 11 and discusses the theory of mobile robots. It gives the mathematical models for the robot and several control designs. The chapters of Part II are as follows:

- Chapter 8 discusses different modeling approaches and control objectives.
- Chapter 9 gives details of the mathematical modeling.

- Chapter 10 provides details of the control algorithm for several different controllers.

- Chapter 11 describes how to simulate the robotic car using *matrix laboratory* (MATLAB), a powerful mathematics program.

The appendices of the book provide the information needed to build a robotic car. This includes hardware schematics, parts lists, and vendor information.

Now, let's build a robot!

Part

1

Hardware Implementation

Overall System Structure

In this chapter, we give an overview of the FLASH car's architecture. We discuss the system structure that was used and describe each subsystem and sensor used. This chapter provides a high-level view of how the FLASH car works and the chapter acts as a precursor to the rest of Part I. The detailed workings of the hardware and software are described in Chapters 4, 5, 6, and 7.

Before we begin with the description of the car, it is appropriate to first talk about the construction of the track upon which the car drives. Knowing the environment helps to understand the various devices that are used for sensing. The FLASH lab is shown in Fig. 2.1. As evident in the picture, the track has a black surface with a white line down the middle. The car travels down the center of the track while straddling this white line. In addition, portions of the track are fitted with magnets underneath. Since the car can detect the presence of both the white line and the magnets, either can take precedence and allow the car to follow different paths. The operation of the different sensors is discussed later in this chapter.

As the FLASH acronym indicates, one important feature of the car is that it be low cost. To keep the cost down, the car is designed with as many low-cost, off-the-shelf components as possible. As such, the basic structure of the vehicle is standard *remote control* (RC) model car equipment. The components that make up the basic structure of the car are given in Table 2.1.

The components in Table 2.1 are used "as is" except for the electric motor. Because the motor is manufactured for RC car racing, it is capable of traveling up to 35 mph. However, the FLASH car is used to simulate real driving situations and thus does not need to exceed 10 mph. Therefore, the motor is rewound with 100 turns of 30 *American Wire Gauge* (AWG) wire to reduce its top speed.

Other than the components in Table 2.1, the rest of the electronics on the FLASH car were designed in house specifically for use in this project. The only

FIGURE 2.1   A portion of the FLASH lab track.

TABLE 2.1   Standard RC Components Used on the FLASH Car

Component	Manufacturer
Legends 1:10 scale model car kit	Bolink
Standard servo	Futaba
P2k Pro Stock Motor	Trinity
7.2V NiMH battery	RadioShack

exception is the *digital signal processor* (DSP), which is purchased as a development kit (described in the following pages).

The car's overall architecture is shown in Fig. 2.2. The system can be broken down into four hierarchical levels. First, several sensors provide information about the car and its location in its environment. This information is sent to the second level, a high-level processor that acts as the "brain" of the car by interpreting the information and deciding how to move the car. This main processor sends commands to the third level, a low-level processor that translates the commands into drive signals. The drive signals are sent to the last level, the motor and servo, that drive and steer the car, respectively.

Each of the sensors shown at the top of Fig. 2.2 provides different information to the main processor. The *infrared* (IR) and magnetic sensors indicate where the path is located directly beneath the car. The camera provides path

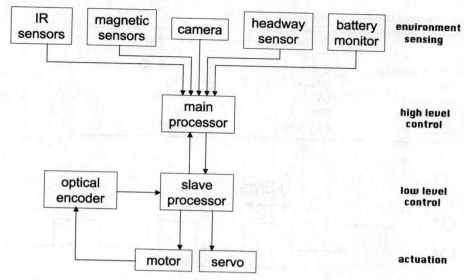

FIGURE 2.2 Overview of the car's hardware architecture.

information about the area in front of the car. The IR, magnetic, and camera sensors can be used for lateral control to determine how to steer the car. The headway sensor can use either infrared or ultrasound technology to indicate how close an object is to the front of the car. The information is used for automatic cruise control or headway control. Finally, the battery monitor gives information about how much power remains. Knowing the power level helps determine when the car needs recharging. There is another sensor that does not communicate directly with the main processor. This is the optical encoder that measures the speed of the wheels to determine how fast the car is traveling. This information is needed to maintain a constant speed on hills. The optical encoder is connected to the slave processor that sends the car's speed to the main processor. The operation of the optical encoder is described later in this chapter. Table 2.2 lists all the sensors used on the vehicle.

Actuation

This section discusses the components used for actuation.

DC motor

The DC motor drives the rear wheels of the vehicle. A DC motor converts DC current into torque. The more current that is sent through the windings in the

TABLE 2.2   Sensors Used on the FLASH Car

Component	Model	Manufacturer
Reflective object sensor	QRD1114	Fairchild Semiconductor
Hall effect sensor	HAL506UA-E	Micronas
CMOS image sensor	OV7610	OmniVision
Infrared range finder	GP2D12	Sharp
Optical encoder	HEDS-I00	US Digital

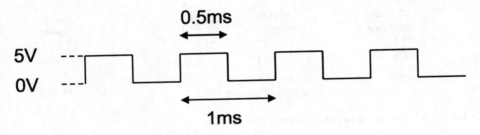

FIGURE 2.3   PWM control signal with a 50-percent duty cycle.

motor, the more torque it produces and the faster the shaft turns. Since current is proportional to voltage, the shaft speed is proportional to the voltage across the motor. However, motors are typically controlled by digital electronics that only output 0V and 5V. However, if you apply 0V or 5V to the motor, it will only run at two speeds. To control the motor using a digital signal, a *pulse width modulation* (PWM) signal is used. PWM signals take advantage of the fact that DC motors cannot respond as quickly as digital electronics can operate. So, if the motor is given a signal, as in Fig. 2.3, which is changing between 0V and 5V at a speed of 2 kHz with a 50-percent duty cycle (defined as the ratio of the high time to the period of the signal), the motor cannot possibly turn on and off so quickly. The result is that the motor acts as if it is receiving 2.5V. In general, the voltage seen by the motor is the average over one period of the PWM signal.

Another issue with motor control through digital electronics is that such circuitry cannot deliver the current a motor needs. The impedance of a DC motor is only a few ohms. If a digital processing chip is connected directly to a motor, the chip will surely be damaged. Special interface circuitry known as an H-bridge is usually used to convert the low-power digital signal to a higher power one and also isolate the motor from the more sensitive digital electronics. A typical H-bridge is shown in Fig. 2.4. This circuit works by turning on opposite transistors simultaneously; for example, transistors A and D are turned on while B and C are turned off. In such a configuration, the current flows through

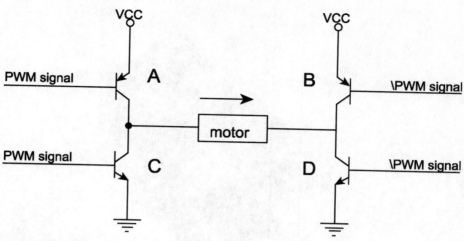

FIGURE 2.4   An H-bridge circuit used to drive a motor.

the motor in the direction of the arrow. If the situation is reversed (B and C are on while A and D are off), the current flows in the other direction, causing the motor to turn opposite of the previous direction. The transistors are turned on and off by the digital electronics. In designing an H-bridge controller, it is important that transistors A and C are not turned on simultaneously, as that would cause a short between the power and the ground. Likewise, transistors B and D should not be turned on at the same time.

As mentioned above, the DC motor in use is the Trinity P2k Pro Stock Motor shown in Fig. 2.4. To reduce the top-end speed, the motor is rewound with 100 turns of 30AWG wire. This also increases the motor resistance and reduces the amount of current needed to drive the car.

Servomotor

The purpose of the servomotor is to provide steering for the car. A servomotor is a DC motor with its own circuitry included. This circuitry acts as a feedback position controller. When a PWM signal is sent to the servo, the motor shaft goes to a particular angular position instead of turning continuously. The standard PWM signal for servos is shown in Fig. 2.6. It has a pulse width between 1 and 2 ms and a period of 10 to 20 ms. A pulse width of 1.5 ms makes the wheels point straight ahead, while 1 ms turns them 45 degrees left and 2 ms turns them 45 degrees right.

The servo used on the car is the Futaba S3003 standard servo. It is shown in Fig. 2.7 and its characteristics are shown in Table 2.3. This is a basic low-cost servo that is used in many RC hobby cars and boats.

FIGURE 2.5   The DC motor used for driving the car's rear wheels.

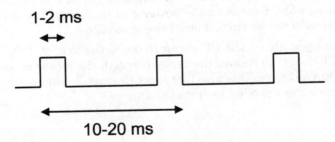

FIGURE 2.6   Format for the standard servo PWM control signal.

Sensors

This section discusses different types of sensors.

Infrared (IR)

The FLASH car utilizes the Fairchild Semiconductor QRD1114 IR emitter/detector pairs to locate the white line on the black road surface. The sensor is shown in Fig. 2.8(a) and its internal circuitry is shown in Fig. 2.8(b). Sending a

FIGURE 2.7 The servo used for steering the car.

TABLE 2.3 Technical Information for the Servo

Device name	S3003
Manufacturer	Futaba
Speed @ 4.8V	0.23 sec/60°
Torque @ 4.8V	44.3 oz/in
Speed @ 6V	0.19 sec/60°
Torque @ 6V	56.8 oz/in
Size	1.6" × 0.8" × 1.4"
Weight	1.31 oz

current into pin 3 forward-biases the emitter, causing it to send out IR light with a wavelength of 940 nm. The detector is located next to the emitter in the same package. The detector consists of a phototransistor that is turned on in the presence of IR light. When the sensor is over the white line, the light from the emitter is reflected back and is seen by the detector, which turns on the transistor. When the sensor is over the black line, no light is reflected. Nothing is detected and the transistor remains closed. The specifications for the IR sensors are given in Table 2.4.

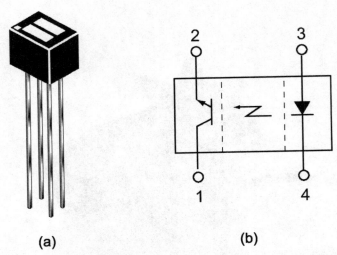

(a)   (b)

FIGURE 2.8 (*a*) The IR sensor package. (*b*) The IR sensor internal circuitry. (Images courtesy Fairchild Semiconductor Corporation.)

TABLE 2.4   Technical Information for the IR Sensors

Device name	QRD1114
Manufacturer	Fairchild Semiconductor
Emitter forward current	50 mA maximum
Emitter forward voltage	1.7V
Peak emission wavelength	940 nm
Sensor dark current	100 nA
Collector emitter saturation voltage	0.4V maximum
Collector current	1 mA minimum
Package size	0.173" × 0.183" × 0.240"

These sensors, while extremely reliable under controlled circumstances, are very susceptible to sunlight and debris on the road surface. More importantly, without major changes in the infrastructure of the roadways today, these sensors will not be viable because they require the line to be underneath the car rather than beside it.

Magnetic

The FLASH car also uses the HAL506UA-E Hall effect sensors from Micronas. The sensor is shown in Fig. 2.9(a). These tiny sensors are used to locate the magnetic line that is beneath the roadway by detecting the presence of a magnetic south pole. Similar to the phototransistor in the IR sensor, a *field effect*

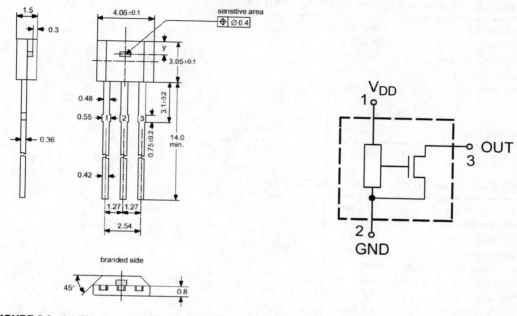

FIGURE 2.9 (*a*) The Hall effect sensor package. (*b*) This Hall effect sensor internal circuitry. (Images courtesy Micronas GmbH, Freiburg, Germany.)

transistor (FET) device is turned on in the presence of 5.5 mT of a magnetic field and turned off when the field strength falls below 3.5 mT. The internal circuit is shown in Fig. 2.9(b). This particular Hall effect sensor is simple to use and requires a minimum of connections; only power, ground, and a pull-up resistor are needed for operation. Table 2.5 summarizes the specifications for these Hall effect devices.

Unlike the IR sensors, the Hall effect sensors do not detect a visible line and are not susceptible to interference from ambient light or roadway debris. However, they require the presence of a magnetic field that is not embedded in most public roadways.

Vision

The vision system includes the use of a digital camera that is mounted, facing forward, on the front of the vehicle. This camera provides the images of the roadway ahead of the car. The camera's specifications are given in Table 2.6.

The primary reason for choosing a DSP platform is the processing power required to rapidly process visual data. (DSPs are described later in this chapter.) Of all the systems, the visual system is perhaps the most intuitively easy

TABLE 2.5   Technical Information for the Hall Effect Sensors

Device name	HAL506UA-E
Manufacturer	Micronas
Operating voltage	3.8–24V
Supply current	3 mA
Switching type	Unipolar
B_{on}	5.5 mT
B_{off}	3.5 mT
Package size	4.06 mm $\times$ 1.5 mm $\times$ 3.05 mm

TABLE 2.6   Technical Information for the Digital Camera

Device name	OV7610
Manufacturer	OmniVision
Operating voltage	5V
Operating current	40 mA
Max. frames per second	60
Array size	644 $\times$ 484 pixels
Pixel size	8.4 µm $\times$ 8.4 µm
Effective image area	5.4 mm $\times$ 4 mm
Communication interface	I^2C package size 40 mm $\times$ 28 mm

to understand because it so closely resembles the way that humans drive automobiles today. A digital color camera in conjunction with the DSP comprises the vision system utilized in this project. The DSP interfaces with the camera via a *direct memory access* (DMA) channel and the interconnection bus to capture video frames of the road ahead. The camera is an OmniVision model OV7630 color *complimentary metal oxide semiconductor* (CMOS) camera.

Although it is a color camera, it can be used as a black and white camera, which is more than sufficient for use in the FLASH project. Future enhancements to the vision system may include the incorporation of color analysis to recognize special or hazardous situations, such as recognizing orange construction cones or special hazard signs. Once the camera has uploaded a frame of image data to the DSP, the processor uses edge-detection and prediction algorithms to determine the curvature of the upcoming road. Thus, the camera data can be incorporated with the other sensor data to allow the vehicle to adjust its speed, and to travel more efficiently down straight stretches of road and through turns.

Headway sensing

The Sharp GP2D12 range finder is used to determine whether an object is directly in front of the car. This device consists of an infrared emitter/detector pair and additional circuitry as shown in Fig. 2.10. However, it is different from the IR sensors described previously; the IR light emitted from this device is modulated at 40 kHz. That is, the light coming out turns on and off 40,000 times per second. This makes the sensor less susceptible to interference from ambient light. Ambient light is usually not modulated at 40 kHz; it produces a constant intensity on the detector. The detector in this device has a bandpass filter that blocks all frequencies other than 40 kHz from getting through, so only the IR light originating from its emitter is seen. The additional circuitry needed to accomplish the modulation and filtering makes the device much bulkier than the IR sensor discussed earlier in the chapter. Because of its size, the GP2D12 is not suitable for mounting in an array beneath the car looking down; however, it is perfectly suitable for mounting a single device on the front of the car looking forward for objects.

The closer an object is to the sensor, the more powerful the light that is received at the detector. The output of the device is an analog voltage that increases as an object moves closer. Fig. 2.11 shows the output voltage versus the

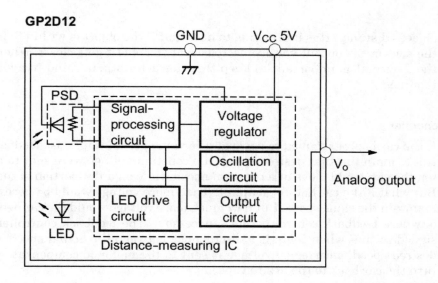

FIGURE 2.10   Internal block diagram of the GP2D12 IR range finder (image courtesy Sharp Corporation).

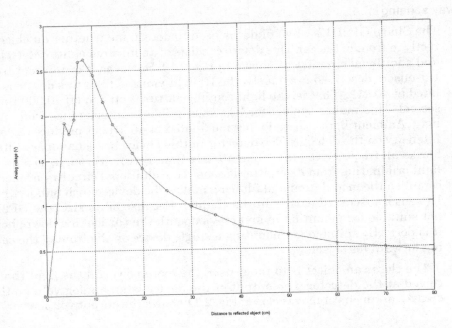

FIGURE 2.11   The GP2D12 output voltage characteristic curve (image courtesy Sharp Corporation).

object's distance. It is interesting to note that if the object is within 8 inches of the sensor, the output voltage decreases at the object gets closer. When using the device, it is important to keep this characteristic in mind to avoid false readings.

Optical encoder

If the car is programmed to drive at a certain speed, it must use feedback control to maintain that desired speed. If a constant voltage were sent to the motor, the car would move at a constant speed while on a flat portion of the track. But what if the car had to go uphill? Then more voltage would be necessary to maintain the same speed. But how much more voltage would the car need? This is where feedback control becomes necessary. The feedback controller is designed so that when there is a discrepancy between the actual speed and the desired speed, a corrected voltage is sent to the motor to compensate and return the car back to the desired speed.

An optical encoder is utilized to measure the speed of the wheels. An optical disk with slits like the one shown in Fig. 2.12(a) is connected to the car's rear axle. An optical emitter/detector pair is placed on either side of the disk. The

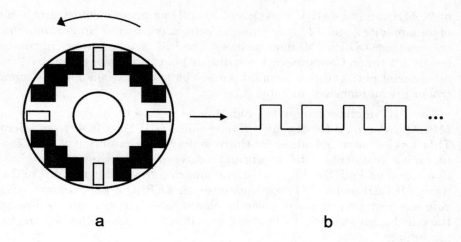

a   **b**

FIGURE 2.12   (*a*) The optical disk placed on the rear axle of the car (*b*) The output signal generated by the encoder.

slits in the disk allow light from the emitter to reach the detector, causing the detector to turn on and off as the disk rotates. Thus, the output of the encoder is a square wave whose frequency is proportional to the angular velocity of the rear wheels as in Fig. 2.12(b). The optical encoder used on the FLASH vehicle is the U.S. Digital HEDS-9140-I00 that has 512 slits in the optical disk.

Processors

On the FLASH car, low-level vehicle control is done with a microcontroller, and high-level control is accomplished with a DSP. The microcontroller receives commands from the DSP and interprets those commands to generate signals that directly control the steering servo and motor.

PIC microcontroller

The microcontroller chosen was the PIC16F874 from Microchip. This particular model was chosen because it is ideal for performing the down-conversion from the DSP to the motor and servo. This architecture allows the DSP to perform high-level control while the microcontroller interfaces directly with the actuators.

The PIC16F874 is a 40-pin, high-performance *reduced instruction set computer* (RISC) processor that has a total of 35 executing instructions. It operates at 20 MHz and has 4K × 14 words of FLASH program memory, 192 × 8 bytes of data memory, and 128 × 8 bytes of *electrically erasable programmable read-only memory* (EEPROM) data memory. The PIC also has two 8-bit timers and one 16-bit timer. Communication with peripherals is done through a synchronous serial port and five parallel ports. The specifications for this microcontroller are summarized in Table 2.7.

The PIC microcontroller is programmed in assembly language using the MPLAB *Integrated Development Environment* (IDE) available from Microchip. This development tool allows for the assembly and simulation of code. The simulator is convenient for debugging and code verification. Once the code has been assembled and a HEX file created, the program can be downloaded to the chip using MPLAB and a PIC programmer or an EEPROM programmer. After the chip has been programmed, it can be placed into a circuit, and upon power-up, the code begins executing. Chapter 6 goes into detail about the PIC and its programming.

Digital signal processor (DSP)

The microprocessor used in this application is the TMS320C6711C by Texas Instruments. This is a specific type of microprocessor known as DSP. A DSP was chosen over a microcontroller for the car because DSPs are well suited for numerically intensive applications such as this one. Additionally, C compilers are available for the TI family of DSPs, thus eliminating the burden of writing assembly code.

The C6711 DSP is a 32-bit floating-point device operating at 150 MHz. There is 64K of internal memory available as well as access to 64M × 32 of external RAM.

TABLE 2.7   Technical Information for the PIC Microcontroller

Device name	PIC16F874-20/P
Manufacturer	Microchip
Operating voltage range	2.0–5.5V
Operating frequency	DC-20 MHz
FLASH program memory	4 K × 14-bit words
Data memory	192 bytes
EEPROM data memory	128 bytes
Interrupts	14
I/O ports	5
Timers	3
Capture/compare modules	2
Instruction set	35 instructions
Package type	40-pin DIP

The chip also has a built-in boot loader, so programs can be loaded from EEPROM and run on the DSP. Additional peripherals include a serial channel, 16 DMA channels, and 2 timers. Table 2.8 summarizes the features of the C6711 DSP.

One advantage of the C6711 is that it comes with a *DSP Starter Kit* (DSK). The DSK enables the user to connect the DSP to the parallel port on a PC and download code using a *graphical user interface* (GUI). This interface allows the programmer to step through the code on the DSP and check the values of registers or variables while debugging. Although appropriate for development, this is not practical in the final system, as the program must be started using the PC and then disconnected. Fortunately, the DSK has on-board EEPROM so that the code can be loaded onto the programmable chip. On power-up, the DSP loads the program from this chip into its own memory and begins execution. Details of the DSP programming are given in Chapter 7.

Power

The following section discusses power.

Batteries

Those who have used RC cars are probably familiar with the car batteries. RC car batteries come in two varieties: *nickel cadmium* (NiCad) and *nickel metal hydride* (NiMH).

NiCad batteries are usually used for racing because they can produce a lot of power. However, they can only sustain such intense levels for a few minutes before recharging becomes necessary. In addition, they must be completely discharged before being recharged so that they do not form "memory lock." If they are charged starting with a high voltage level, NiCad batteries will not charge to their full potential. This reduces the battery's useful lifetime and they will need to be replaced more often. These characteristics make NiCad batteries well suited for racing situations, but not for the FLASH lab.

TABLE 2.8   Technical Information for the C6711 DSP

Device name	TMS320C6711C
Manufacturer	Texas Instruments
Processor type	32-bit floating point
Operating voltage	3.3V
Operating frequency	150 MHz
Cycle time	5 ns
Serial ports	1
DMA channels	16
Timers	2

The alternative battery type, NiMH, is much more suitable for the laboratory environment. Although NiMH batteries produce less power than NiCad, high power is not needed if the cars are not running at high speeds. Since less current is used, they discharge at a much slower rate. This means that the cars have a longer runtime, which is useful when testing the car's performance. Also, NiMH batteries do not have the same tendency as NiCad to form "memory lock." The time to recharge a battery is reduced because a discharge is not necessary each time. The batteries used in the FLASH lab are 7.2V NiMH 3000 mAH and are available from many hobby stores. One such battery is shown in Fig. 2.13.

A third type of battery that is common in many consumer electronic devices, but not RC cars, is lithium ion. These batteries have the potential for a longer runtime than the batteries currently in use. However, a few issues prevented us from choosing this type. First, lithium ion batteries are heavier than NiMH when compared volt for volt. Thus, the gain from increased runtime may be offset by the fact that the car must work harder to propel the increased weight. Second, lithium ion batteries can explode when not charged correctly. Since there is an automatic recharging system in the FLASH lab and the batteries are charged while in the car, safety is a concern. Therefore, lithium ion batteries are not used on the FLASH car.

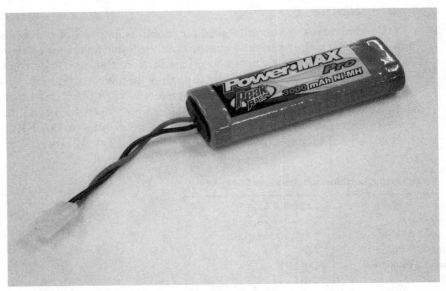

FIGURE 2.13   A NiMH battery used on the FLASH car.

Automatic recharging

With all of the electronics included on the car, the NiMH battery provides 1 to 1.5 hours of runtime. And because this car was partially designed as part of a museum exhibit, it is desirable to have the car run with as little intervention from the staff as possible. Thus, the car is capable of self-monitoring and automatic recharging. An overview of this subsystem's operation is given here.

The flowchart for the recharging algorithm is shown in Fig. 2.14. While the car is operating normally, the DSP keeps track of the battery voltage using a

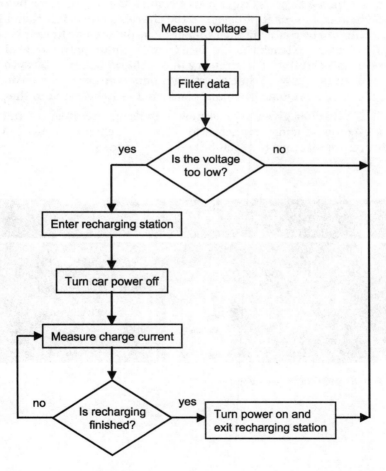

FIGURE 2.14   Flowchart for the recharging system.

monitoring circuit. (The circuitry is described in detail in the next chapter.) After reading this data, it must be filtered so that noise in the sensor does not falsely cause the car to go into recharge mode. If the battery voltage is sufficient, the car continues along the path; however, if the voltage is below threshold, the car goes into recharge mode.

In normal running mode, the car uses the IR sensors as the primary input for following a path. However, in recharge mode, the car switches to the magnetic sensors as the primary sensors. The charging station is located off the track (see Fig. 2.15) and consists of one bay for each car that is running on the track. The pull-off for each bay is denoted by a perpendicular line across the track. Each car is assigned a bay, and the car determines the bay number by counting the perpendicular lines as it pulls out of the charging station. It then knows which bay to pull into by counting the number of perpendicular lines it sees. There are magnets under the track that deviate from the main path and lead the car into the recharging station. The circuitry in the charging bay detects the car's presence and shuts it down. When charging is done, the car is turned on. It exits the charging station, rejoins the main track, and switches back to the IR sensors.

This chapter has given an overview of the car's architecture and the components used for sensing and actuation. The next chapter shows how all of this hardware fits together to become the FLASH car.

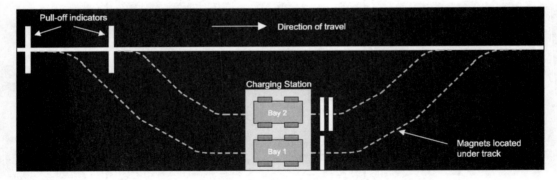

FIGURE 2.15 The charging station configuration.

3

Construction

This chapter gives step-by-step instructions on how the FLASH car is assembled. The parts needed are listed in Appendix A. It should be noted that *printed circuit boards* (PCBs), shown here, have been fabricated for the FLASH car. In particular, one PCB was designed so that it fits directly into the chassis in place of the battery tray.

First, the motor must be rewound so that its top speed is reduced, and thus its ability to produce more torque is increased. Pull out the brushes and remove the mounting screws as shown in Fig. 3.1. Remove the motor cap and pull the armature out of the motor housing. See Fig. 3.2.

On the armature, carefully lift up the tab holding the wire in place as in Fig. 3.3. Do not damage the brass axle, as this is where the brushes make contact. Cut and unwind the wire to completely remove it from the armature. Scrape the insulation off the end of the 30AWG wire and solder it to one of the tabs. See Fig. 3.4.

Wrap the wire around one leg of the armature 100 times. Where the wire passes over the next tab, scrape off the insulation and solder the wire to the tab as in Fig. 3.5. Repeat this for each leg of the armature. On the last leg, cut the wire to length so that it just passes over the starting tab. Complete the winding by soldering the end of the wire to this tab. See Fig. 3.6. Now reassemble the motor as in Fig. 3.7. This is the motor that will be installed into the chassis.

Next, the parts from the Bolink Legends kit must be assembled. Although the kit comes with assembly instructions, the assembly has to be modified to accommodate the additional electronics on the car. Some of the instructions here may vary slightly from those given with the kit because of differences in the parts that are supplied.

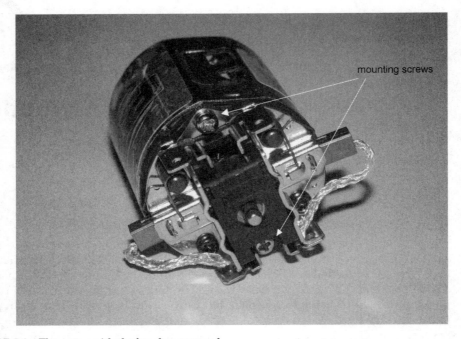

FIGURE 3.1   The motor with the brushes removed.

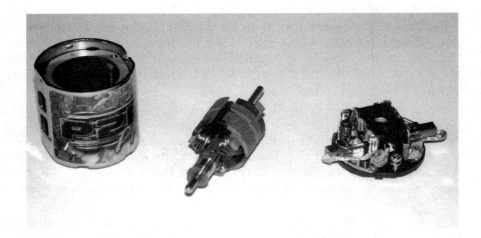

FIGURE 3.2   The disassembled motor.

FIGURE 3.3 Lift the tab holding the motor wire in place.

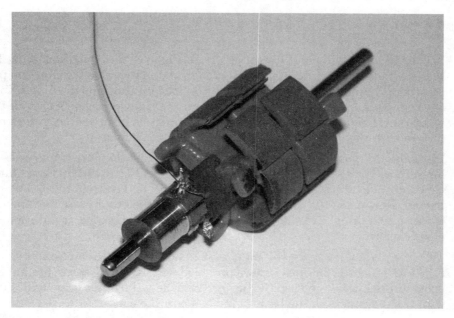

FIGURE 3.4 The starting wire connection.

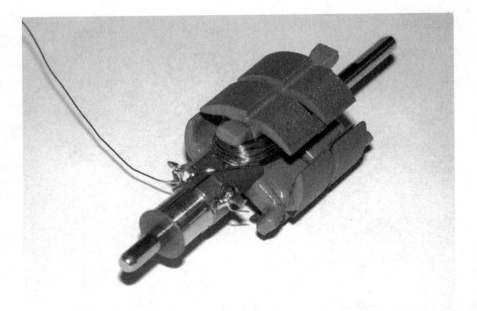

FIGURE 3.5   The wire connection between each armature leg.

The chassis plates and bushings are in Fig. 3.8. The bushings must be inserted where the rear axle passes through the plates. (Notice that the plastic inserts may not be included with the kit.) If the kit contains the plastic inserts, one of them must be modified as shown on the right in Fig. 3.9. Cut off the lip on the short side of one of the inserts. This must be done so that it does not interfere with the optical encoder that will be attached later.

Press the cut plastic insert into one of the plates and then press the brass bushing into the plate using the orientation shown in Fig. 3.10. This orientation is very important because the optical encoder will be attached to the left side of the car. Now press the uncut plastic insert into the other plate and then press the brass bushing into the plate as in Fig. 3.11. Notice the orientation of the plastic insert.

To the plate with the modified plastic insert, attach three posts as shown in Fig. 3.12 using the 3/8-inch screws provided with the kit. Notice that not all the posts in the kit will be used. If they are attached, the other posts will interfere with the placement of the battery underneath the car.

Insert the front axle plate and battery tray into the side plate. Slide on the other side plate, and attach the plate to the three posts using the 3/8-inch screws, as shown in Fig. 3.13. Notice that on the FLASH car, the battery tray

FIGURE 3.6   The rewound armature.

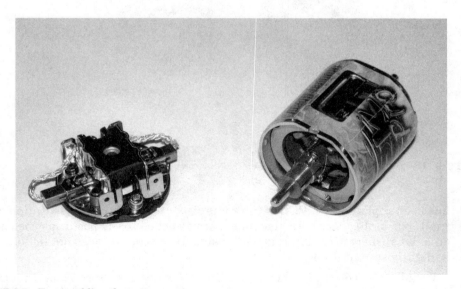

FIGURE 3.7   Reassembling the motor.

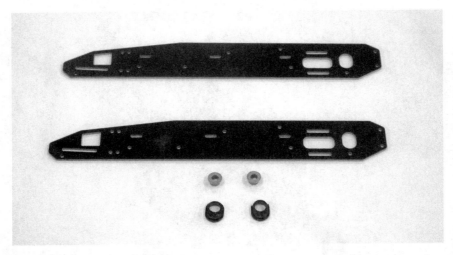

FIGURE 3.8   The chassis plates and bushings.

FIGURE 3.9   The original insert (left) and the modified insert (right).

has been replaced by a PCB of the same size. Also, notice the orientation of the front axle plate. Once both side plates are attached to the posts, lay the chassis on a flat surface to ensure it is not twisted. If the chassis does not lie flat, twist it slightly so that it does.

Install e-clips on the toothed end of the stub axles and press them through the steering block, using the orientation shown in Fig. 3.14. Notice that the kit may

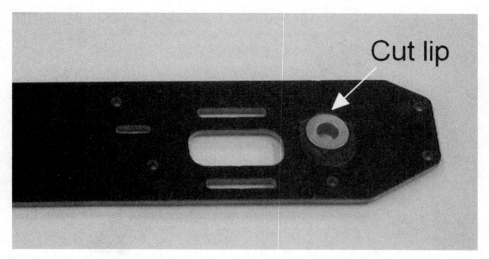

FIGURE 3.10 The location and orientation of the cut insert.

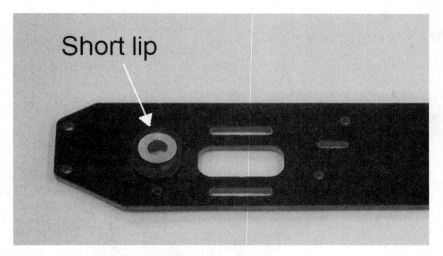

FIGURE 3.11 The location and orientation of the uncut insert.

contain stub axles without teeth. Install each steering block as shown in Fig. 3.15. Use the provided lock nut to secure the steering block from underneath.

Insert ball bearings into every other hole in the outside row of the differential gear. See Fig. 3.16. Put the rear axle through the side plates with the orientation shown in Fig. 3.17. Install the components of the diff according to the instructions supplied with the kit. It should resemble Fig. 3.18.

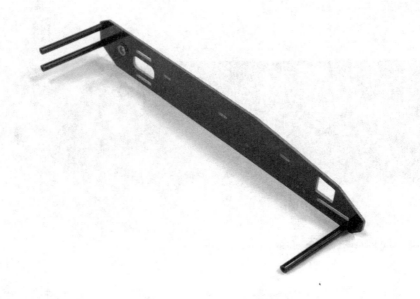

FIGURE 3.12   Chassis post attachment.

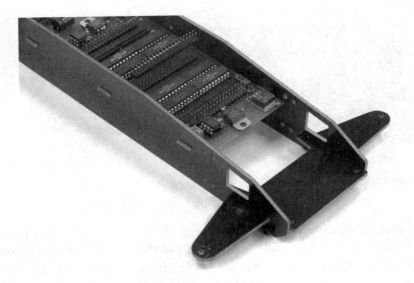

FIGURE 3.13   The proper orientation of the front axle plate.

FIGURE 3.14   Stub axle insertion into the steering block.

FIGURE 3.15   The assembled front steering axle.

FIGURE 3.16 The diff gear with the ball bearings installed.

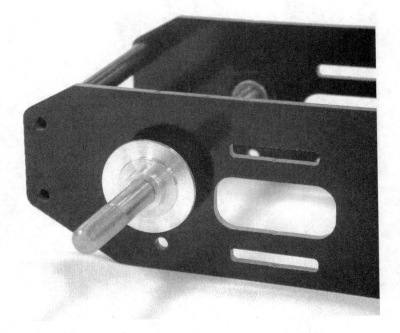

FIGURE 3.17 The orientation of the rear axle.

FIGURE 3.18 The assembled differential.

On the left side of the rear axle, install the two nylon washers and the optical encoder wheel. See Fig. 3.19. The two washers are necessary so that the encoder wheel passes through the encoder module. Leave a small amount of room between the encoder wheel and the chassis so that there is some play in the rear axle. Notice that the nylon spacer included with the kit is not used and has been replaced with the optical encoder wheel.

Next, using hot glue, install the optical encoder module with the orientation shown in Fig. 3.20. Make sure that the module fully encloses the optical wheel so that a signal is output. You can verify this by connecting 5V to pin 5 and ground to pin 1 (pin 1 is to the left in the figure). Observe pin 3 on an oscilloscope. When you spin the rear axle, a square wave should be output from pin 3. After the optical wheel, install the setscrew hub. See Fig. 3.21.

Solder the motor wires to the motor tabs and install the pinion gear on the rewound motor. On the FLASH car, the gear included with the kit is not used and has been replaced with a 14T gear. Using two hex screws and two washers, set the motor so that the pinion gear and the diff gear mesh. There should be a small amount of play between them. See Fig. 3.22.

Remove the mounting bracket from the servo and insert the long steering links into the star-shaped servo saver as shown. The servo saver arms that do not hold the steering links must be cut off so that they do not interfere with the wiring, which will be added later. Place the servo so that it sits against the side chassis plate. The steering links should pass through the square holes in the

FIGURE 3.19   The location of the optical encoder wheel on the axle.

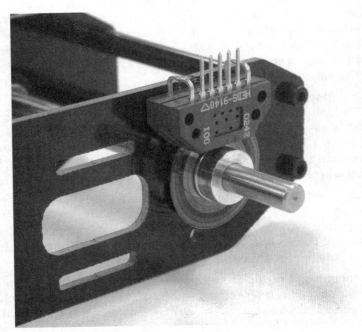

FIGURE 3.20   The encoder module must fully enclose the optical wheel.

FIGURE 3.21   The setscrew hub.

FIGURE 3.22   The motor gear should mesh with the diff gear.

side chassis plates. Secure the servo using the tape provided with the kit as in Fig. 3.23. Install the short steering links, front wheels, and rear wheels according to the instructions provided with the kit.

Next, place and secure the *infrared* (IR)/magnetic sensor circuit boards below the front and rear axles. Connect them to the existing circuit board using ribbon cable as in Fig. 3.24. Place the headway sensor in the center of the servo and secure with hot glue as shown in Fig. 3.25.

FIGURE 3.23   The location of the servo.

FIGURE 3.24 The front IR/magnetic sensor board.

FIGURE 3.25 Mount the headway sensor using hot glue.

At this point the FLASH car should look like Fig. 3.26.

The last two circuit boards to be installed on the car are the PIC board (Fig. 3.27) and the DSP board (Fig. 3.28). The DSP board is an off-the-shelf component from Texas Instruments, whereas the other circuit boards were designed as part of the FLASH project.

FIGURE 3.26   The partially assembled FLASH car.

FIGURE 3.27   The PIC board.

FIGURE 3.28   The DSP board.

Once the PIC board is placed on the car, the wires for the motor and optical encoder must be attached as shown in Fig. 3.29. The wires for the servo and headway sensor are attached as in Fig. 3.30.

The battery is installed underneath the car. The connector is soldered in line, with a switch and a fuse, to the bottom side of the battery tray board. A small *central processing unit* (CPU) fan is glued to the left side of the car to provide cooling for the H-bridge. See Fig. 3.31.

And finally, the completed FLASH car is shown in Fig. 3.32. In the following four chapters, we discuss in detail the car's components and programming.

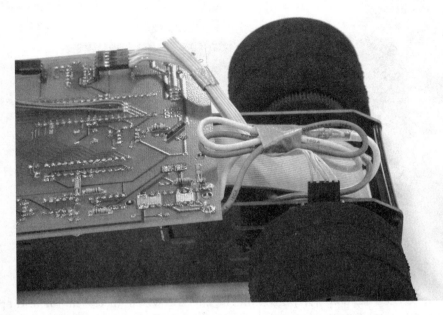

FIGURE 3.29 The wires connecting the motor and optical encoder to the PIC board.

FIGURE 3.30 The wires connecting the servo and headway sensor to the PIC board.

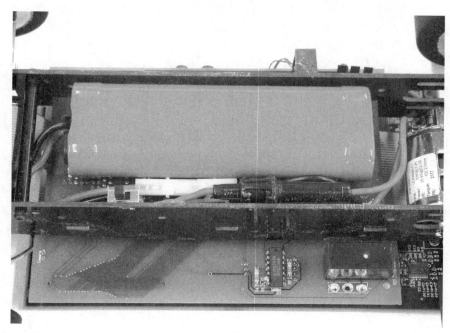

FIGURE 3.31   The battery and CPU fan are mounted underneath the car.

FIGURE 3.32   The completed FLASH car.

Environment Sensing

Now that we have an overall view of the robot's design, we turn to the details. In this chapter, we discuss how the robot senses its environment. We also describe how the robot monitors its own parameters, such as speed and battery voltage, so that such data can also be used for feedback control. The information that the FLASH car acquires for feedback control includes the lateral displacement from the center line, the distance from objects in front of it, its speed, and its power level. Although the previous chapter gave an overview of these monitoring systems and described the sensors, this chapter describes the circuitry surrounding those sensors in more detail.

Lateral Displacement

The objective of the FLASH car is to follow a line. As was seen in Fig. 2.1, the track has a black surface with a white line down the middle. In addition, magnets are mounted below the track surface that act as a secondary line. The car can choose to follow either the white line or the magnets (this is discussed in the car's programming in Chapter 7, "Microprocessor Control").

The FLASH car contains arrays of *infrared* (IR) and magnetic sensors, also known as *Hall effect devices* (HED), to help determine where the car is located with respect to the desired path. See Fig. 4.1. As shown in Fig. 4.2, these sensors are mounted underneath the front and rear of the car. The sensors look straight down at the roadway to determine where the desired path is underneath the car. The thickness of the line and the spacing of the sensors are such that two are activated at any time. The sensor configuration is shown in Fig. 4.3. Note that the rear sensors are directly between the rear wheels. This is done so that the reading from the rear sensors gives the lateral placement along the rear axle directly. There are 12 IR sensors and 12 magnetic sensors in each array; this

FIGURE 4.1   The sensor arrays on the bottom of the car: the magnetic sensors (above) and the IR sensors (below).

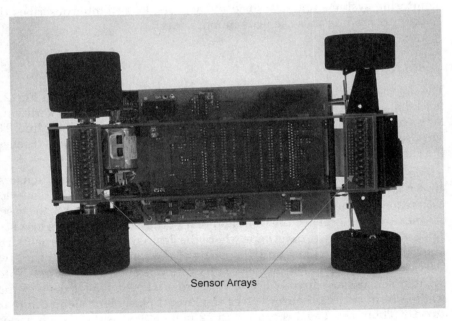

FIGURE 4.2   The location of the sensor arrays on the bottom of the car.

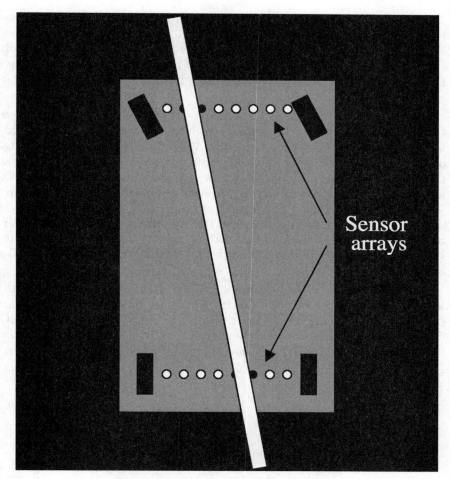

FIGURE 4.3   Only the sensors located above the white line are activated.

number provides good resolution. However, the *digital signal processor* (DSP) software is configured so that any number of sensors can be used. In theory, the number of sensors is limited only by the width of the DSP's data bus. In this section, we describe how the car senses both the white and magnetic lines.

IR circuitry

The IR sensor consists of an IR *light-emitting diode* (LED) and a phototransistor together in a single package. When current flows through the LED, IR light is emitted. The phototransistor is activated in the presence of IR light.

The light acts in the same way as the base current would in a regular transistor. When the sensor is over the black part of the track, the transistor is off. When the IR light is reflected by the white line and sensed by the phototransistor, it is turned on. However, the amount that the transistor is activated, and thus the amount of current running through it, depends on the amount of light acting on it.

The circuit for the IR sensor is shown in Fig. 4.4. The resistor on the left determines how much current flows through the emitter. The value should be selected to send approximately 10 to 20 mA through the LED with a voltage drop of 1.7V. In the FLASH car implementation, this resistor is a potentiometer so the intensity of the emitted light can be adjusted. The detector side of the pair is open collector so the resistor on the right serves to pull up the output to 5V. When no light is detected, the transistor is off and V_{out} is pulled up to 5V. As the amount of light increases, the current flowing through the transistor increases, making V_{out} decrease. If the light is strong enough, the transistor saturates and V_{out} becomes close to 0V. This circuit is repeated 12 times to form the IR sensor array.

Now, we only care about whether a line is present or not. So we are interested in a digital signal, on or off, 5V or 0V. There are different ways to accomplish this. On the FLASH car, V_{out} is sent through a digital buffer, the SN74HCT244. This device translates any voltage above 2V to 5V and any voltage below 2V to 0V. In addition, the buffer can be enabled or disabled, which is useful when

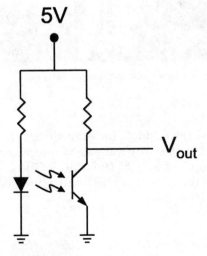

FIGURE 4.4 The circuit application for the IR sensor.

more than one device must communicate over the same set of wires. This multiple access data bus is described later in this chapter.

Magnetic circuitry

The circuitry for the magnetic sensor is simpler than the circuitry for the IR sensors. It is only necessary to connect 5V and ground to the power pins and connect a pull-up resistor to the output. This is shown in Fig. 4.5. When in the presence of a magnetic field, the device turns on and the output voltage is 0V. Otherwise, the output is pulled up to 5V by the resistor. Unlike the IR sensors mentioned previously, the magnetic sensor is a digital device; its output is either 0V or 5V, but never in between. However, the output of the device is sent to a digital buffer. This is done so that the magnetic array data can travel over the single data bus. As with the IR, this circuit is repeated 12 times to form the magnetic sensor array.

Headway Measurement

The FLASH car has also been programmed with automatic cruise control. The car will maintain its desired speed unless it comes upon a slower-moving object. In that case, the car will slow down to maintain a minimum distance between itself and the object in front of it. So, we must have a sensor that will measure the distance of objects in front of the car. This device is the IR range finder.

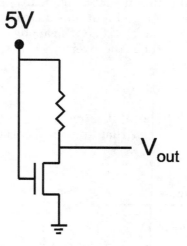

FIGURE 4.5   The circuit application for the magnetic sensor.

IR range finder

The overall operation of the IR range finder was described in the previous chapter. Like the IR sensors used for lateral displacement, the IR range finder consists of an emitter and detector in the same package. However, the light emitted is modulated at 40 kHz, and the detector has a bandpass filter that only lets that signal through. The output of the sensor is an analog voltage that is related to the distance the reflected light has traveled.

Recall that the IR sensors used for lateral displacement also output an analog voltage that was then sent through a buffer to digitize the signal . We could do this because we were only interested in whether the white line was present or not. However, for headway sensing, we want to know more than whether the object is in front of the car. We also want to know *how far* in front of the car the object is. For this, we need the analog signal.

Headway circuitry

Because we will need to digitize the signal with an *analog-to-digital* (A/D) converter (described later in this chapter), as well as deal with quite noisy analog signals, the signal from the sensor must go through some conditioning circuitry. This circuitry cleans up the signal by filtering out some of the noise and buffers the signal to ensure that there is no problem when the signal arrives at the A/D converter.

The circuit shown in Fig. 4.6 consists of a low-pass filter with a cutoff frequency of about 1.6 kHz. Experimentally, it was found that this cutoff frequency was able to block disruptive noise while not noticeably slowing the measurements. The output of the filter is sent through an *operational amplifier* (opamp) buffer. Such a buffer has a high input impedance so that the range finder does not have to source more current than it's capable of delivering. The output of the buffer is then sent to the A/D converter.

Speed Measurement

To maintain a constant speed, it is necessary to measure the speed of the wheels. Recall from Chapter 2 that an optical encoder placed on the rear axle

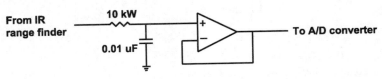

FIGURE 4.6   The circuit used to buffer the output of the IR range finder.

is used for this measurement. This device outputs a square wave whose frequency is proportional to the rotational velocity of the rear wheels. Some circuitry is required to interpret the output of the encoder.

Encoder circuitry

For the particular optical encoder used on the FLASH car, the HEDS-9140-I00 from US Digital, there are actually two square waves that are output. They are denoted as Channel A and Channel B. Although they are the same frequency, they are 90 degrees out of phase with one another as shown in Fig. 4.7. Using these two signals, you can determine which direction the wheel is turning. (This is a helpful bit of information if you want to know if the car is driving forward or backward!)

A D flip-flop is used to interpret this information. A D flip-flop is a synchronous device in which the logic level on the input pin, D, gets transferred to the output pin, Q, on the rising edge of a clock signal. In addition, there are asynchronous preset and clear pins that put the output to a high or low level, respectively. The typical function of a D flip-flop is given in Table 4.1.

The two channels of the optical encoder are connected to two D flip-flops as shown in Fig. 4.8. The $\overline{PRE}$ inputs are not shown because they are not used and are connected to 5V. The resulting signal, for the case when the car is going forward, is shown in Fig. 4.9. The signal from channel A leads channel B by 90 degrees. When channel A has a rising edge, the logic level of channel B is transferred to the output of the top D flip-flop, the reverse signal. When the car is going forward, channel B is always low when there is a rising edge on channel A. So in this case, the reverse signal is always low.

When channel B has a rising edge, the logic level of channel A is transferred to the output of the bottom D flip-flop, and in the forward case, this level is always high. Then, when channel A has a falling edge, the output of the flip-flop is cleared. The output changes immediately and does not wait for a rising edge of the clock.

Looking at the circuit in Fig. 4.8, the top D flip-flop outputs pulses when the car is going in reverse; conversely, the bottom D flip-flop outputs pulses when

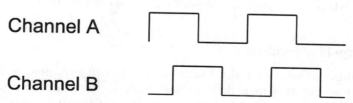

Channel A

Channel B

FIGURE 4.7 The two signals output by the optical encoder are 90 degrees out of phase.

TABLE 4.1   Function Table for a D Flip-Flop

$\overline{PRE}$	$\overline{CLR}$	CLK	D	Q	$\overline{Q}$
L	H	x	x	H	L
H	L	x	x	L	H
H	H	↑	H	H	L
H	H	↑	L	L	H

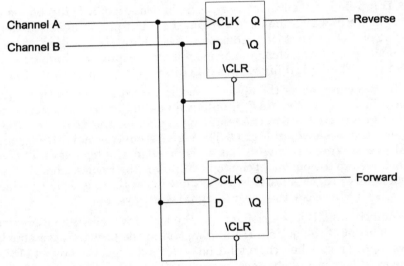

FIGURE 4.8   The circuit connections for decoding the optical encoder output using two D flip-flops.

the car is going forward. These forward and reverse signals are then sent to the PIC microcontroller that counts the pulses to determine the car's speed. The PIC's method for counting is described in the next chapter.

Battery Voltage Measurement

An additional feature of the FLASH car is that it can monitor its own health, or its power level. This is done as part of the automatic recharging system. We have designed the car and track system so that it can operate 24 hours a day on its own. Therefore, it is not necessary for a person to replace discharged cars

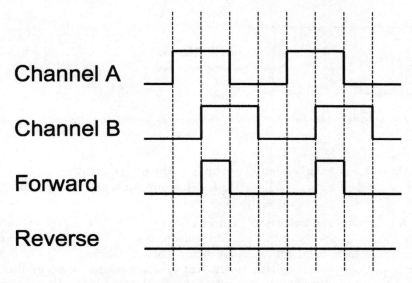

FIGURE 4.9 The output signals from the D flip-flops, when the car is moving forward.

with ones that are fully charged. The cars monitor the battery's voltage and current to determine when a charge is necessary.

The battery can be modeled, as shown in Fig. 4.10, as an ideal voltage source in series with an internal battery resistance, R_{in}. The voltage that is measured at the battery terminals is V_{bat}. To know what the actual battery voltage is, it is necessary to know R_{in}, V_{bat}, and I_{bat}. The values for V_{bat} and I_{bat} can be measured while the car is running (this is described in the following sections). However, R_{in} must be measured offline.

To determine the value of R_{in}, at least two measurements of V_{bat} must be made using two different values for R_L. Then R_{in} is given by

$$R_{in} = \frac{V_{bat1} - V_{bat2}}{\dfrac{V_{bat2}}{R_{L2}} - \dfrac{V_{bat1}}{R_{L1}}}$$

where V_{bat1} and V_{bat2} are the voltages measured when load resistors R_{L1} and R_{L2} are used, respectively. If more measurements are taken using several different values for R_L, the results can be averaged, minimizing the effect of measurement error. The internal resistance is typically very small, on the order of 0.15 ohms.

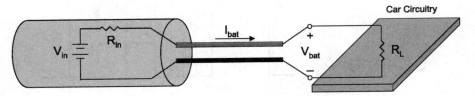

FIGURE 4.10   The battery modeled as an ideal power source in series with the internal battery resistance.

Once the internal resistance is known, the internal voltage of the battery can be calculated as $V_{in} = V_{bat} + R_{in} I_{bat}$. The measurement of V_{bat} and I_{bat} are described in the following sections.

A fully charged battery measures about 8.3V at the battery terminals. The car can run properly until the voltage drops to about 6.5V. The DSP keeps track of the voltage level and decides when the car should pull off the track to recharge. (This is described in Chapter 7.) This means that a digital representation of the voltage must be sent to the DSP. As with the IR range finder, the analog signal must be conditioned and put through an A/D converter.

Voltage-sensing circuitry

The voltage-sensing circuit is shown in Fig. 4.11. This circuit serves two purposes. First, it filters noise and smoothes the signal. The 1 µF capacitors act in conjunction with the resistors to form low-pass filters that eliminate frequencies above about 3 Hz. This eliminates any noise that may be present, as well as voltage spikes that the motor may cause.

Second, the voltage-sensing circuit shifts and amplifies the signal to take advantage of the full range of the A/D converter. The A/D converter can measure voltages between 0V and 5V. However, the battery voltage ranges from about 6.5V to 8.3V. The circuit shown in Fig. 4.11 cuts the battery voltage in half using a voltage divider (lower-left opamp). Then, 3.14V is subtracted out and the result is multiplied by 4.7. This makes the input at the A/D converter between 0.5V and 4.8V, with the value being $V = (\frac{V_{bat}}{2} - 3.14)*4.7$. In the DSP software, the value read in is converted back to the battery voltage using the inverse relationship: $V_{bat} = 2*(\frac{V}{4.7} + 3.14)$.

In the circuit in Fig. 4.11, the reference voltage value of 4.096V was chosen for convenience. This voltage is generated by the A/D converter and is readily available on one of its pins. The reference voltage is put through a voltage divider, resulting in the 3.14V that is subtracted. There are many other voltage reference devices on the market that could be used instead. For example, you may wish to choose another reference value that would change the formula used by the DSP to calculate the battery voltage.

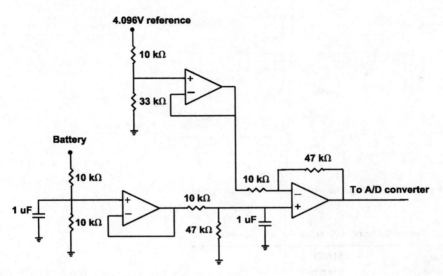

FIGURE 4.11 The circuit for measuring the battery voltage.

Current-sensing circuitry

In addition to monitoring the battery voltage, the car can also monitor the current being drawn from the battery. If too much current is flowing from the battery, something is wrong and the car knows to shut itself down. The circuit used to monitor the battery current is given in Fig. 4.12. This circuit uses a current sense resistor, a device that has very low resistance and can handle high current. It must have low resistance so that there is not too large of a voltage drop, which would interfere with the car's operation. It must also be high power enough to handle up to 5 amps. In this case, the current sense resistor is 0.05 ohms and is connected between the battery's negative terminal and the circuit ground. Therefore, the current coming from the battery is converted to a voltage by the current sense resistor. The voltage is then multiplied by 50 in the first opamp circuit. The second opamp circuit is a low-pass filter with a very low cutoff frequency to eliminate noise from the signal. The voltage that is sent to the A/D converter is given by $V = I_{bat}*0.05*50$. This circuit can measure up to 2 amps before the opamp saturates and the measurements become very inaccurate.

A/D Conversion

The A/D converter used on the FLASH car is the Maxim MAX196, a 6-channel, 12-bit converter that operates off a single 5V supply. Table 4.2 gives the specifications for the chip. This device takes an analog input on any of its 6 channels

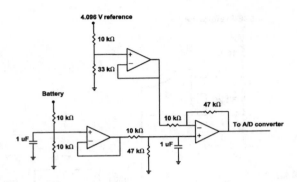

FIGURE 4.12   The circuit for measuring the battery current.

TABLE 4.2   Technical Information for the A/D Converter

Device Name	MAX196
Manufacturer	Maxim
Resolution	12 bits
Linearity	$\frac{1}{2}$ LSB
Operating Voltage	5V
Selectable Input Ranges	0–5V, 0–10V, ± 5V, ± 10V
Analog Input Channels	6
Reference Voltage	4.096V
Conversion Time	6 μs
Sampling Rate	100 ksps

and converts it to a 12-bit number between 0 (representing 0V) and 4,095 (representing 5V). The pinout for the MAX196 is shown in Fig. 4.13.

The DSP reads data from the A/D converter to use in its control algorithms. To talk to one another, the DSP must first configure the A/D so that the A/D knows how to convert the analog signal. The communication that takes place is shown in Fig. 4.14. First, the chip must be enabled by pulling the $\overline{CS}$ to ground. The $\overline{WR}$ is sent a low signal so that the DSP can write to the A/D. The configuration word is then sent to the A/D over the data lines (D0 through D7). This configuration word tells the A/D several things that are detailed in the device's datasheet. It is during this time that the DSP tells the A/D which channel to convert from analog to digital. In our case, there are analog voltages from three different sources: the IR range finder, the battery voltage-sensing circuit, and battery current-sensing circuit.

Once the configuration word has been sent, the $\overline{WR}$ and $\overline{CS}$ signals are sent high to end the write sequence and disable the chip. Next, the A/D performs a conversion on the specified channel. This operation takes about 15 μs. Then the

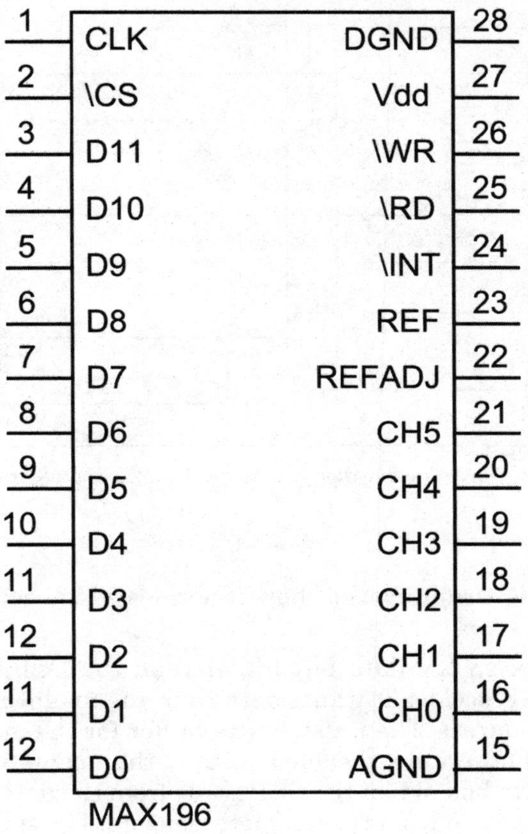

1	CLK	DGND	28
2	\CS	Vdd	27
3	D11	\WR	26
4	D10	\RD	25
5	D9	\INT	24
6	D8	REF	23
7	D7	REFADJ	22
8	D6	CH5	21
9	D5	CH4	20
10	D4	CH3	19
11	D3	CH2	18
12	D2	CH1	17
11	D1	CH0	16
12	D0	AGND	15

MAX196

FIGURE 4.13 The MAX196 A/D converter pinout.

read sequence is begun by pulling the $\overline{CS}$ and $\overline{RD}$ pins to ground. This makes the converted value available on the data pins (D0 through D7) and they are read by the DSP. The details of how the DSP performs these steps are covered in Chapter 7.

Data Bus Interface

Finally, in this chapter we discuss how data is transferred on the FLASH car. Several devices need to communicate with each other, such as the DSP, the PIC, and all the various sensors. It is convenient that they are all connected using the same set of wires (the data bus). However, all the devices cannot send information along the data bus at the same time; they must share it.

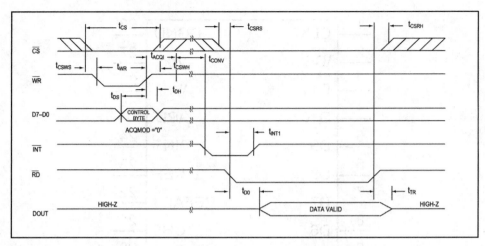

FIGURE 4.14 The timing of an A/D acquisition (image courtesy Maxim Integrated Products, Inc.).

Since the DSP is running the show, it decides which device it wants to talk to and when.

Each device on the data bus (other than the DSP) is given an address. When the DSP decides it wants data from the magnetic sensors, it makes a call to their address. Then, the digital buffer for the magnetic sensors is enabled and all others are disabled, so only the magnetic sensor data has access to the data bus. When the DSP reads from the data bus, it knows it is receiving information from the magnetic sensors. The setup for the interface is shown in Fig. 4.15.

The address decoder shown in Fig. 4.15 interprets the signals from the DSP address bus and determines which device to enable. On the FLASH car, this decoding is implemented by a *programmable logic device* (PLD) called the GAL16V8. This is an electronically erasable logic device that can be programmed to implement a digital circuit. The pinout for the device is shown in Fig. 4.16. There are nine pins that are dedicated inputs and nine pins that can be either inputs or outputs.

To talk to the other devices, the DSP uses the following pin address bus signals: $\overline{CE2}$, $\overline{ARE}$, $\overline{AWE}$, A2, and A3. The PLD converts these into enable/disable signals for each of the devices, as well as $\overline{RD}$ and $\overline{WR}$ signals for the A/D converter and PIC. (The PIC's interface with the data bus is described in the next chapter.) The decode logic is given in Table 4.3. By convention, the enable, read, and write signals are active low, as indicated by the "!" preceding the variable name. The code to implement this logic is as follows:

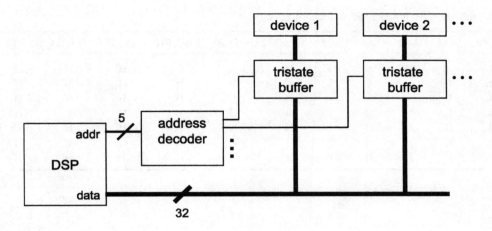

FIGURE 4.15 Interface between the DSP and the peripheral devices.

1	IN0	Vcc	28
2	IN1	I/O8	27
3	IN2	I/O7	26
4	IN3	I/O6	25
5	IN4	I/O5	24
6	IN5	I/O4	23
7	IN6	I/O3	22
8	IN7	I/O2	21
9	IN8	I/O1	20
10	Vss	I/O0	19

GAL16V8

FIGURE 4.16 The programmable logic device pinout.

TABLE 4.3   Logic Implemented on the GAL16V8

!CE2	!ARE	!AWE	A3	A2	!IR_EN	!HED_EN	!PIC_EN	!ADC_EN	!READ	!WRITE
H	x	x	x	x	H	H	H	H	H	H
L	L	x	L	L	L	H	H	H	H	H
L	L	x	L	H	H	L	H	H	H	H
L	L	H	H	L	H	H	L	H	L	H
L	H	L	H	L	H	H	L	H	H	L
L	L	H	H	H	H	H	H	L	L	H
L	H	L	H	H	H	H	H	L	H	L

```
Device G16V8 ;

/* ************** INPUT PINS ********************/
PIN [1..2] = [A3,A2];
PIN 3      = !ARE;
PIN 4      = !AWE;
PIN 5      = !CE2;
/* ************** OUTPUT PINS ********************/
PIN 19 = !IR_EN ; /* Infrared array */
PIN 18 = !HED_EN ; /* Hall effect array */
PIN 17 = !PIC_EN ; /* PIC parallel slave port */
PIN 16 = !ADC_EN ; /* MAX196 parallel ADC */
PIN 15 = !READ ; /* */
PIN 14 = !WRITE ; /* */


FIELD ADDR = [A3,A2];

IR_EN  = CE2 & !A3 & !A2 & ARE; /* 00 */
HED_EN = CE2 & !A3 & A2 & ARE; /* 01 */
PIC_EN = CE2 & A3 & !A2; /* 10 */
ADC_EN = CE2 & A3 & A2; /* 11 */
READ   = CE2 & ARE & (PIC_EN # ADC_EN);
WRITE  = CE2 & AWE & (PIC_EN # ADC_EN);
```

This code must be compiled into HEX format so that it can be programmed into the GAL16V8. Several free programs are available on the Internet to do this. One popular program that is simple to use is from Atmel and is called WinCupl. It is available from www.atmel.com.

If the code given previously is saved to a file with a .pld extension, it can be opened in WinCupl. See Fig. 4.17. To verify that the code performs as expected, an add-on program called WinSim (available with WinCupl) can be used to sim-

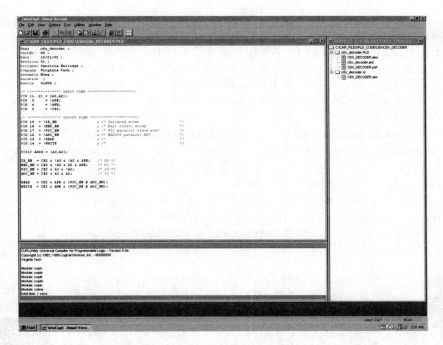

FIGURE 4.17 The WinCupl development environment.

ulate the code with various inputs. First, a simulation file must be created that goes through the various inputs that are to be tested. A possible simulation file for the previous code is given here:

```
ORDER: ADDR, !ARE, !AWE, !CE2, !IR_EN, !HED_EN, !PIC_EN,
!ADC_EN, !READ, !WRITE;

VECTORS:
00111HHHHHH
01111HHHHHH
10111HHHHHH
11111HHHHHH
00010LHHHHH
01010HLHHHH
10010HHLHLH
10100HHLHHL
11010HHHHLLH
11100HHHHLHL
```

Each row under VECTORS contains different inputs that are to be tested, as well as their corresponding outputs. The first five columns give logic levels for the inputs ADDR, !ARE, !AWE, and !CE2. (ADDR contains two bits.) The last six columns give the correct outputs for !IR_EN, !HED_EN, !PIC_EN, !ADC_EN, !READ, and !WRITE. These outputs are checked against the simulation results. This file must be saved with an .si extension so that it can be recognized by WinSim.

To invoke WinSim, go to Run/Device Dependent Simulate. This opens an additional program that shows the results of the simulation. See Fig. 4.18. This generates outputs based on the code you have written and allows you to verify that it is giving the desired results.

Once the code has been verified, it must be compiled into HEX format for downloading to the chip. By going to Run/Device Dependent Compile, WinCupl creates an output file that is compatible with the chip identified in the .pld file

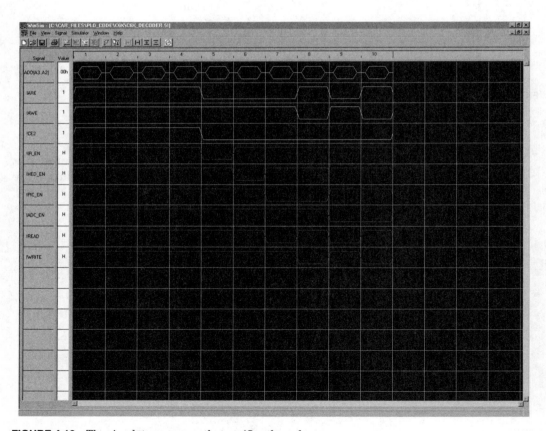

FIGURE 4.18 The simulator program that verifies the code.

(the GAL16V8 in this case). Several output files are created during the compile process. The one that is needed for programming the chip has a .jed extension. This is the file that is downloaded to the chip.

The next step is to program the PLD. This is done using an *electronically erasable programmable read-only memory* (EEPROM) programmer, such as the BK Precision BK844 model shown in Fig. 4.19. EEPROM programmers are available from many electronics suppliers such as Jameco or Digikey, and cost between $200 and $1,000, depending on how many chips are supported. They connect to a PC through the serial or parallel port.

The programmers come with software, so downloading the HEX file to the chip is fairly easy. The software interface for the BK844 is shown in Fig. 4.20. To program a chip, simply insert it into the programmer. From the BK844 window, press the Select button to identify which chip is being used. (In this case, it is the GAL16V8.) Open the .jed file to be downloaded and click on the Auto button. If everything was done correctly, there will be a message indicating that the chip was programmed successfully. Now the PLD is ready for use in the circuit.

One advantage of using the GAL16V8 as an address decoder is that it is programmable. One disadvantage is that it is programmable. The programmability works both for it and against it. On the one hand, the pins and their use can

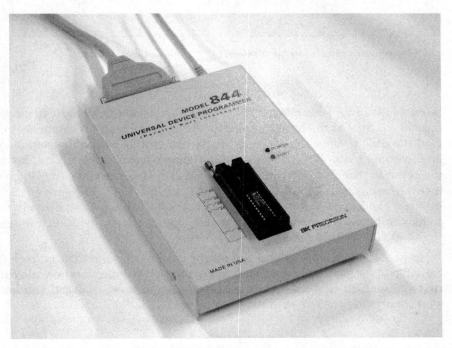

FIGURE 4.19   The BK Precision BK844 EEPROM programmer used for downloading code to the PLD.

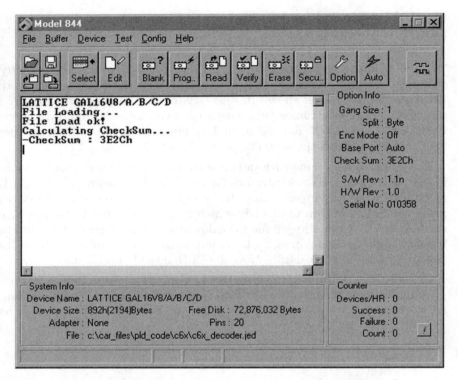

FIGURE 4.20 The software supplied with the BK844 programmer for downloading code.

be customized for your own needs, which allows you to add more devices to the data bus later. On the other hand, this device requires a programmer to download the code to the chip. Although these are readily available from many major electronics suppliers, they are rather expensive. However, if you are willing to surrender the flexibility of the PLD, this address decoder can also be implemented using two 3×8 decoders such as the 74LS138. The pinout and truth table for the 74LS138 are shown in Fig. 4.21 and Table 4.4 respectively. One decoder can be used to generate the enable signals while the other can generate the read and write signals.

Now that we know the details of all the circuitry on the car that collects and transfers data over the data bus, we move on to the details on how the car is actuated. The next chapter describes the motor and servo in detail and how they work to provide the car with movement.

TABLE 4.4   Truth Table for the 74LS138

G1	G2A+G2B	C	B	A	Y0	Y1	Y2	Y3	Y4	Y5	Y6	Y7
x	H	x	x	x	H	H	H	H	H	H	H	H
L	x	x	x	x	H	H	H	H	H	H	H	H
H	L	L	L	L	L	H	H	H	H	H	H	H
H	L	L	L	H	H	L	H	H	H	H	H	H
H	L	L	H	L	H	H	L	H	H	H	H	H
H	L	L	H	H	H	H	H	L	H	H	H	H
H	L	H	L	L	H	H	H	H	L	H	H	H
H	L	H	L	H	H	H	H	H	H	L	H	H
H	L	H	H	L	H	H	H	H	H	H	L	H
H	L	H	H	H	H	H	H	H	H	H	H	L

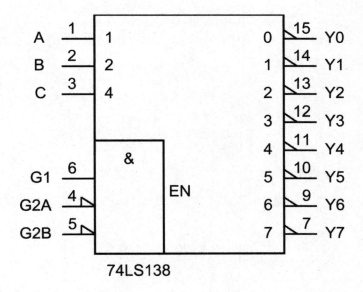

FIGURE 4.21   The pinout for the 74ls138 decoder.

DC Motors and Servos

This chapter describes the action of the DC motor that is used in the car for driving, as well as the servo that is used for steering. The chapter provides the foundations and basic constructions for the motor and servo drivers.

Magnetic Forces on a Moving Charge

When a charged particle travels with a velocity in a magnetic field, it experiences a force. This effect has been observed experimentally and is the basis on which many electromagnetic devices, including motors, operate. In fact, there is a quantitative relationship between the velocity of the particle, the charge of the particle, the magnetic field, and the force experienced by the particle. As shown in Fig. 5.1 and Fig. 5.2, when the angle between the velocity vector of the positive charge and the magnetic field vector is 90 degrees, the force experienced by the particle of positive charge q is given by

$$|F| = q|v|\|B\| \tag{5.1}$$

Equation 5.1 shows the magnitude of the force equal to the product of the charge q, the magnitude of the velocity $|v|$, and the magnitude of the magnetic field, $|B|$. On the other hand, if the angle counterclockwise from the velocity vector to the field vector is θ, then the force experienced by the positive charge is

$$|F| = q|v|\|B\|\sin\theta \tag{5.2}$$

In both cases, the direction of the force is perpendicular to both velocity and field vectors, and is given by the direction of the motion of a screw when you twist a screw in the direction of increasing angle θ. In vector notation, the force can be written in terms of a vector cross product.

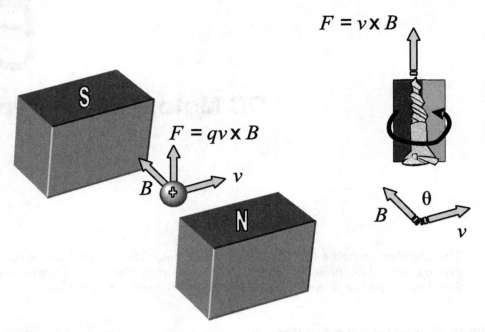

FIGURE 5.1   The force experienced by a charged particle.

$$F = qv \times B \tag{5.3}$$

This vector notation simply states that the magnitude of the force is given by equation (5.2). If you turn a screw in the direction of increasing θ, the direction of F is the same as the screw, and perpendicular to both v and B.

Magnetic Forces on a Conductor

Now let us consider a conductor with current flowing through it. If you consider an infinitesimal length of the conductor dL, it has a charge dq traveling at a velocity v. The force on the positive charge is given by

$$F = dqv \times B \tag{5.4}$$

We can write the velocity as

$$v = \frac{dL}{dt} \tag{5.5}$$

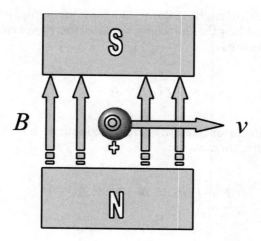

$$\odot \quad F = qv \times B$$
Direction of force is outward

FIGURE 5.2   The force experienced by a charged particle (plane view).

We can write the force as

$$F = dq\frac{dL}{dt} \times B \tag{5.6}$$

This can also be written as

$$F = \frac{dq}{dt} dL \times B \tag{5.7}$$

We can write the current in the conductor as

$$I = \frac{dq}{dt} \tag{5.8}$$

This expression combined with (5.7) gives the force experienced by a section of the conductor of length dL as

$$F = IdL \times B \tag{5.9}$$

In Fig. 5.3, the entire length of the conductor is straight, giving the same force for every differential length. Therefore, the total force experienced by the entire length of the conductor is

$$F = I \int dL \times B \qquad (5.10)$$

If the field is uniform, or constant, and the conductor has the same angle with the field everywhere, then we get

$$F = I \int dL \times B = IL \times B \qquad (5.11)$$

Magnetic Forces on a Loop

The DC motor operation is based on the principle that a loop of a conductor that is carrying current experiences a torque that makes it rotate. This principle is shown in Fig. 5.4.

When the current flows in the loop from the positive terminal of the battery to the negative terminal, in the direction as shown in the figure, the conductor

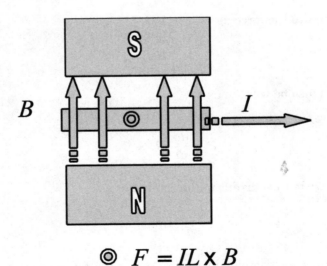

$$\odot \quad F = IL \times B$$

Direction of force is outward

FIGURE 5.3   The force experienced by a current-carrying conductor (plane view).

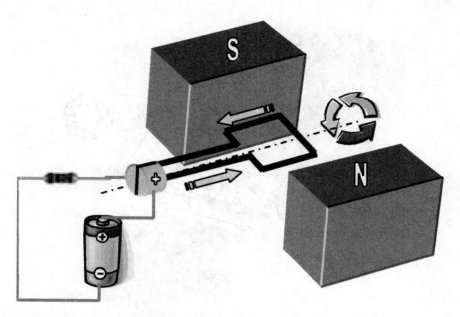

FIGURE 5.4   A current loop for a DC motor.

experiences forces on its various parts due to the magnetic field, which is due to the magnets, which results in a torque that causes the conductor to rotate about its axis. The details of the torque production for a loop are shown in Fig. 5.5.

Looking at Fig. 5.5 and applying equation (5.11) to find forces on the four sides of the current loop, we see that two forces, F_2 and F_4, face each other and do not contribute anything to the rotating torque about the axis of rotation for the loop. However, the forces F_1 and F_3 both contribute to the torque. These two forces have the same magnitude and are given by

$$F_1 = F_3 = BIa \tag{5.12}$$

The torque due to force F_1 is

$$T_1 = \frac{1}{2} BIab \sin \theta \tag{5.13}$$

The torque due to F_3 is the same in magnitude and direction as the one due to F_1. Therefore, the total torque produced is

$$\tau = BIab\sin\theta \tag{5.14}$$

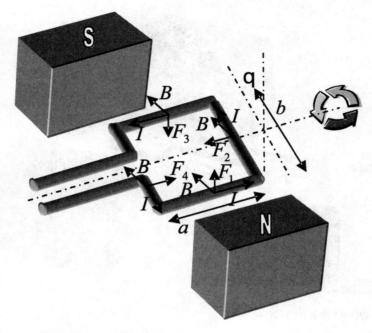

FIGURE 5.5   Torque on a current loop for a DC motor.

This can also be written as

$$\tau = \mu \times B \tag{5.15}$$

where

$$\mu = IA \tag{5.16}$$

The symbol A denotes the area of the loop given by

$$A = ab \tag{5.17}$$

The direction of the vector $\mu = IA$ is given by the right-hand rule. It is perpendicular to the plane of the current loop and is given in the direction of movement of a screw, if you rotate the screw in the direction of the current in the loop.

We can also increase the torque value by having multiple turns of the current loop. If we have n number of turns in a loop, we will have

$$\mu = nIA \tag{5.18}$$

Commutation

In Fig. 5.5, we see how a torque is produced on a current loop in a magnetic field. When this loop rotates by half of a full rotation, the current should face the same direction as the opposite part of the conductor. This is shown in Fig. 5.6.

In the figure, we see that the part of the conductor that had current flowing in one direction has its current direction reversed as it flows across the commutator (the disc that shows the positive sign on one side in the figure). The loop is connected to the commutator via brushes. This allows the torque to be in the same direction.

Torque flows in one direction because it is a function of the angle of rotation of the current loop. To make the torque independent of the angle, we can have multiple current loops. In Fig. 5.7, we see two loops as an example.

By having multiple loops, we can make the torque independent of the angle. For a motor, we obtain a linear relation between the current and torque. This is written as

$$\tau = K_{torque}I \tag{5.19}$$

where K_{torque} is the torque constant.

DC Motor Power Control

When we connect a battery to a DC motor, the motor starts rotating until it reaches a steady-state speed. We can connect a battery to a motor with a switch,

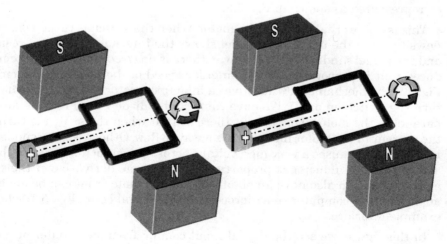

FIGURE 5.6 Commutation.

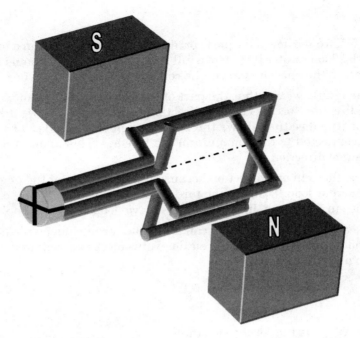

FIGURE 5.7   Multiple loops.

which is shown in Fig. 5.8. When we turn the switch on, the current flows
through the motor and rotates. When the switch is turned off, no current flows,
which causes the motor to stop. This circuit provides an on-off control and can
be represented as shown in Fig. 5.9.

This is an internal model of the motor. When the switch is turned on, current
flows through the motor. The model shows the internal (armature) resistance
and the model's inductance. Moreover, there is an *electromotive force* (emf) produced that is proportional to the angular speed of the motor. In the circuit of
Fig. 5.9, the mechanical switch gives a human operator the ability to manually
turn the switch on or off. We have connected a diode in the reverse direction,
parallel to the motor. This is kept there so that when the switch is turned off,
the current in the motor inductor has a way to flow. Otherwise, the sudden drop
in the current causes a very high voltage across the inductor, because the voltage across any inductor is proportional to the time derivative of the voltage
across it. We can also have an on-off control of a motor that can be enabled by
an output of a computer (microprocessor or a digital chip). Fig. 5.10 shows an
example interface.

In this figure, we see the digital input coming from a computer or a digital
chip. When the input is high, the transistor turns on. That causes the current

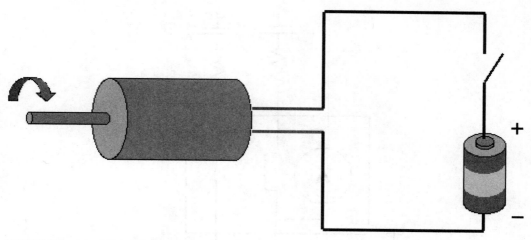

FIGURE 5.8   On-off motor control.

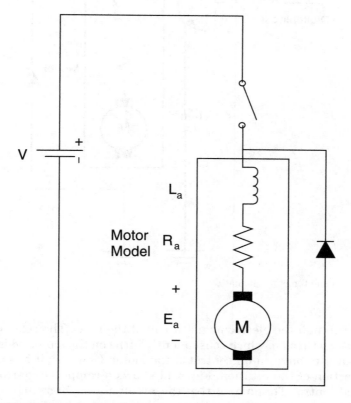

FIGURE 5.9   On-off motor control schematic.

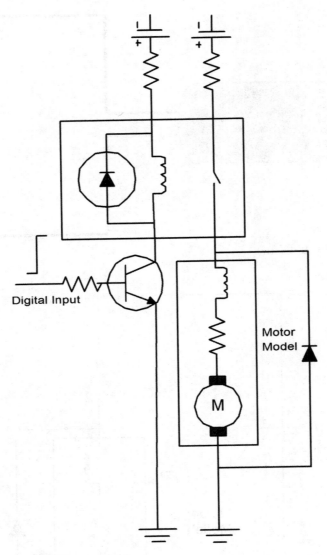

FIGURE 5.10   Digital on-off motor schematic.

to flow through the electromagnetic coil of the relay. The relay coil makes the
relay switch turn on, which consequently turns on the motor. This circuit allows
a computer to have on-off control of the motor. However, it has no control over
the direction of the rotation. Fig. 5.11 shows a computer-controlled, on-off di-
rectional control. That means that the circuit allows for switching the motor on
and off, and also controls the direction of rotation.

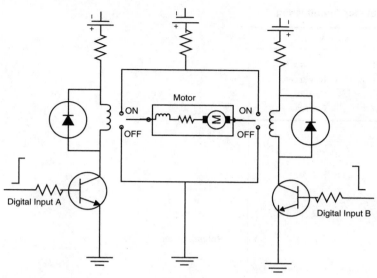

FIGURE 5.11   Digital on-off and direction control schematic.

We see in the schematic that when the digital input A is on and the digital input B is off, the current in the motor flows from left to right. However, if the digital input A is off and the digital input B is on, the current in the motor flows from right to left. When the digital input A is on and the digital input B is also on, there is no current flowing through the motor, and the motor stops due to friction. Similarly, when the digital input A is off and the digital input B is also off, there is no current flowing through the motor, and the motor stops due to friction. These four actions and their effects are summarized in Table 5.1.

Now, we would like to control the speed of the motor. This can be achieved by controlling the current flowing through the motor, since current flow is related to the torque produced in the motor. The voltage applied to the motor controls the amount of current flowing through the motor. One way to control the applied voltage is by using a potentiometer, which is shown in Fig. 5.12. We can also control speed using a transistor, as shown in Fig. 5.13.

In both these schemes, we must increase the resistance in the circuit, thereby reducing the current, in order to reduce the speed of the motor rotation. However, the problem with this method is the power loss in the resistor as well as through the transistor. The power loss through the resistor is proportional to the product of the square of its current and resistance. Let us now understand the power loss through a transistor. Consider Fig. 5.14, which shows a *negative-positive-negative* (NPN) transistor. The analysis for this transistor is more or less valid for all transistors.

TABLE 5.1   Inputs for Circuit in Fig. 5.11

A	B	Motor
On	Off	Forward
Off	On	Reverse
On	On	Brake
Off	Off	Brake

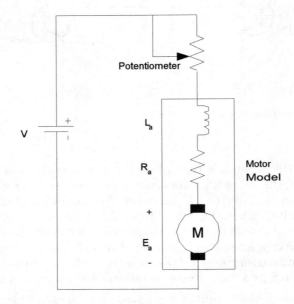

FIGURE 5.12   Potentiometer-controlled speed control.

When the current flowing into the base of the transistor is zero, the current into the collector is also zero. This is called the cut-off region. In this mode, the power loss in the transistor is zero. This is because the power loss is the product of the voltage across the collector and the emitter and the current flowing from collector to the emitter. Because the current flowing in the cutoff region is zero, there is no transistor power loss. The idealized model of this operation is shown in Fig. 5.15.

When the base current is so high that the maximum collector current possible is flowing, we say that the transistor is operating in the saturation region.

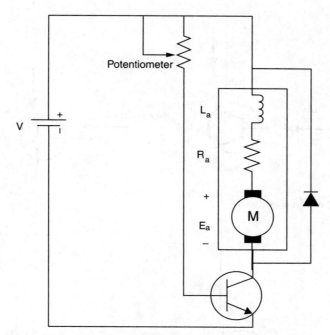

FIGURE 5.13 Potentiometer-controlled speed control using a transistor.

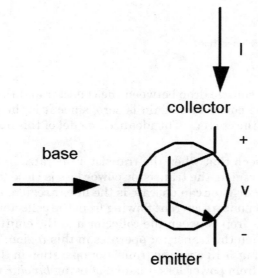

FIGURE 5.14 NPN transistor.

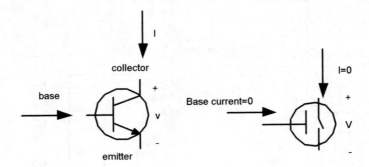

FIGURE 5.15   Cutoff region operation.

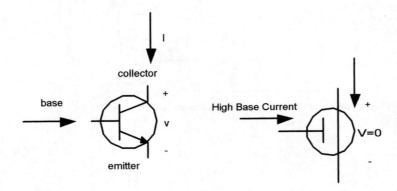

FIGURE 5.16   Saturation region operation.

In this mode, the voltage drop between the collector and the emitter is negligible. Therefore, the power loss again is zero, since it is the product of the zero voltage drop and the current. The idealized model of this operation is shown in Fig. 5.16.

Now, we have seen that when the transistor operates in the cutoff region or in the saturation region, the transistor power loss is close to zero. The third region in which a transistor can operate is the linear region. In that region of operation, there is a nonzero current flowing from the collector to the emitter, and a nonzero voltage drop between the collector and the emitter. Therefore, there is a power loss when the transistor operates in this region. The setup for speed control shown in Fig. 5.13 uses the transistor operation in the linear region and therefore suffers from power loss. Instead of using *bipolar junction transistors* (BJTs), *metal-oxide-semiconductor field effect transistors* (MOSFETs) are used in many applications because of their handling and manufacturing characteristics. These are voltage-controlled devices, as compared to BJTs that are current-controlled devices. In BJTs, the amount of current flowing into the base dictates what region (cutoff, linear, or saturation) the transistor operates in. In MOSFETs, the transistor can be controlled and operated in either the cutoff region, triode or ohmic region, analog of the linear region in BJTs, or the saturation region. We can also make MOSFETs operate as a switch by controlling the voltage between the gate and the source. This is shown in Fig. 5.17. Therefore, a much better way to achieve different speeds from a motor is to provide it with a different amount of current by switching the transistor on and off quickly. This is called a *pulse width modulation* (PWM). This is shown in Fig. 5.18.

The PWM signal is a voltage signal at the gate of the transistor. The signal is a periodic signal. Each time period is divided into two consecutive sections. One is the on time and the other is the off time. Let us use the following variables:

Time period: T

On time within a period: T_{on}

Off time within a period: T_{off}

Duty cycle: On time within a period: $\frac{T_{on}}{T}$

When the duty cycle is 1 (or, when expressed as a percentage, 100 percent), the voltage across the motor is V_M. When the duty cycle is 0, the voltage across

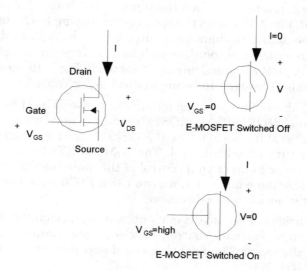

FIGURE 5.17   MOSFET as a switch.

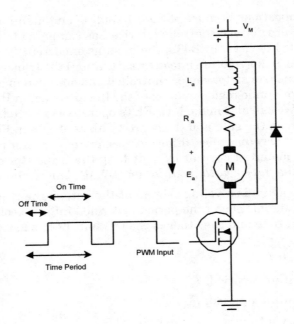

FIGURE 5.18   PWM speed control.

the motor is also zero, because the transistor is off and no current flows through the motor. Using averaged dynamics, we can approximate the voltage across the motor to be equal to dV_M when the input to the transistor is a PWM signal of duty cycle d. Fig. 5.18 shows the speed control using PWM, but the circuit does not have the ability to change the direction of the motor rotation. We need to use PWM to control the speed, as well as to change the direction of the current flowing through the motor. The circuit shown in Fig. 5.19 does just that. The operation of this H-bridge can be understood using Table 5.2.

The table shows the operation for different cases. The H-bridge used here has two *p-channel MOSFETs* (P-MOSFETs) on top and two *n-channel MOSFETs* (N-MOSFETs) at the bottom. The P-MOSFETs turn on when the input is a logic 0, and they turn off with logic 1. The N-MOSFETs behave the opposite. The table shows how bidirectional control of the motor can be obtained. Moreover, instead of keeping enable at 1, we can use a PWM signal on the enable signal to control the speed in any direction.

Some H-bridges also have a current-sensing mechanism via a resistor. This allows us torque feedback control, since the motor torque is proportional to the current in the motor. The following is a description of a commercial H-bridge product LMD18200.

TABLE 5.2   Operation of Circuit in Fig. 5.19

Direction L	Direction R	Enable	A	B	C	D	Operation
1	0	1	Off	On	On	Off	Reverse
0	1	1	On	Off	Off	On	Forward
1	1	1	Off	Off	On	On	Braking
0	0	1	On	On	Off	Off	Braking
X	X	0	Off	Off	Off	Off	Free

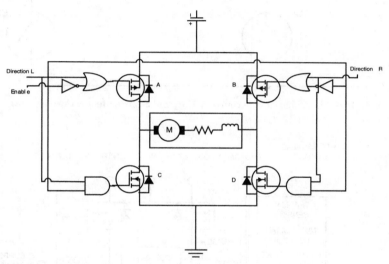

FIGURE 5.19   H-bridge circuit.

LMD18200 H-Bridge

The LMD18200 chip uses the basic H-bridge configuration shown in Fig. 5.20. More details and features are shown in Fig. 5.21. There are three main inputs: direction, brake, and PWM input signals. The control logic truth table for the bridge is given in Table 5.3.

There are two ways to operate this H-bridge. The first way is the locked antiphase. In this method, the PWM signal is connected to 5V and the brake signal is connected to the ground. The actual PWM signal is fed into the direction input pin. A duty cycle of 50 percent gives no motion. A duty cycle greater than 50 percent gives forward rotation, and less than 50 percent gives backward rotation. The amount greater than or less than 50 percent controls the speed also. This is shown in Fig. 5.22.

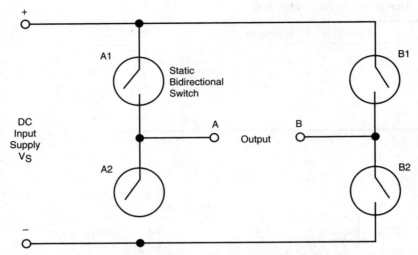

FIGURE 5.20   LMD18200 H-bridge (image courtesy National Semiconductor Corporation).

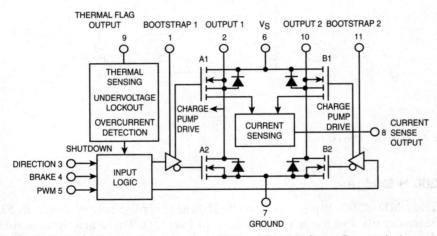

FIGURE 5.21   LMD18200 H-bridge (image courtesy National Semiconductor Corporation).

The other way is to use sign/magnitude control, where the PWM signal is fed through the actual PWM pin, and the direction pin controls the direction of rotation. This is shown in Fig. 5.23.

Figure 5.24 shows an example connection where the H-bridge is controlled via LM629, which is a *proportional-integral-derivative* (PID) controller. The PID controller gets input from an encoder for feedback and is being controlled by a microprocessor.

TABLE 5.3 Operation of the LMD18200 H-bridge

PWM	Dir	Brake	Active output drivers
H	H	L	A1, B2
H	L	L	A2, B1
L	X	L	A1, B1
H	H	H	A1, B1
H	L	H	A2, B2
L	X	H	None

H-Bridge for the FLASH Car

The H-bridge we use for the FLASH car is TPIC0108B. We use this H-bridge because the LMD18200 cannot work with the lower voltage requirements of the car, whereas TPIC0108B can. The functional input/output description of the chip is shown in Table 5.4. The functional block diagram of the chip is given in Fig. 5.25, and the description of the various pins on the chip is given in Fig. 5.26. A typical connection for an application of this H-bridge with a DC motor is shown in Fig. 5.27.

DC Motor Dynamics

In this section, we will develop a mathematical model of the DC motor. We have already seen that the torque produced in a DC motor is proportional to the current applied. Moreover, a back emf produced is proportional to the angular velocity of the motor. These two terms couple the electrical and mechanical parts of the motor operation. Let us derive the dynamics of a motor shown in Fig. 5.28.

When voltage is applied across the motor, the current goes through the armature inductance and resistance of the motor. When we apply external voltage to the motor, current starts flowing in the circuit. This current causes the motor to rotate. We know that when a current-carrying conductor experiences a magnetic field, it consequently experiences a force. The opposite is also true: A moving conductor in a magnetic field produces an emf. Like torque, the emf will be proportional to the angular velocity as well as the angle of the moving coil. However, because of multiple loops in the field, the emf will just be the product of a constant and the angular velocity; it will be independent of the coil angle.

By applying Kirchoff's voltage law in the circuit, we get

$$V = iR_a + L_a\frac{di}{dt} + E_a \qquad (5.20)$$

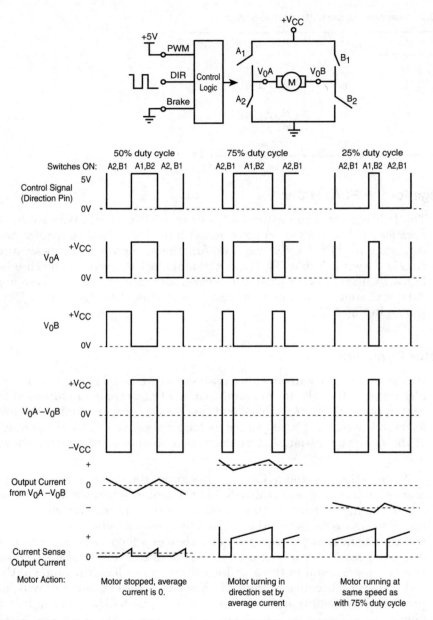

FIGURE 5.22   Locked antiphase control (image courtesy National Semiconductor Corporation).

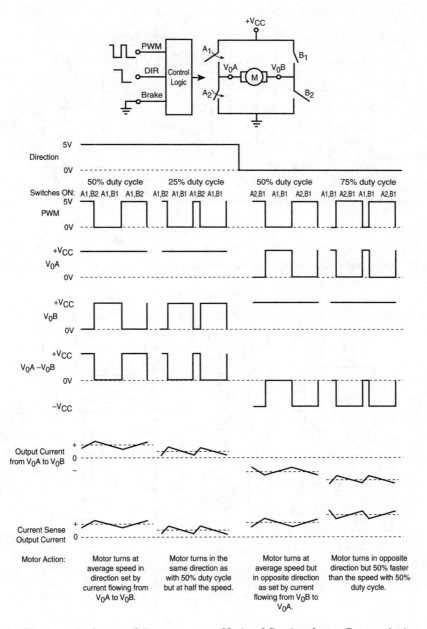

FIGURE 5.23 Sign/magnitude control (image courtesy National Semiconductor Corporation).

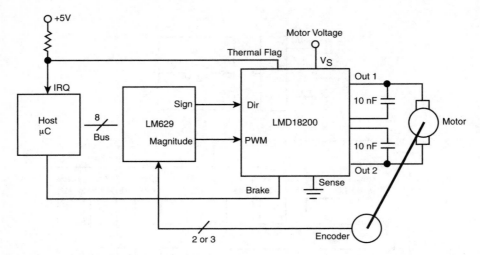

FIGURE 5.24   An example motor control application (image courtesy National Semiconductor Corporation).

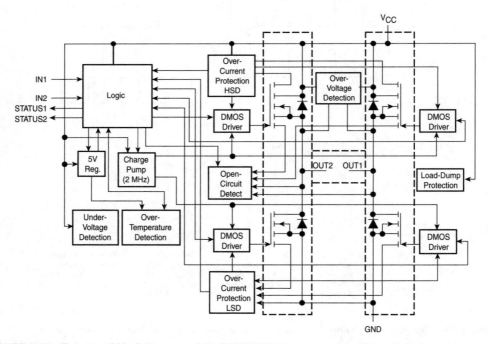

FIGURE 5.25   Functional block diagram of the TPIC0108B (image courtesy Texas Instruments).

TERMINAL NAME	NO.	I/O	DESCRIPTION
GND	7, 9, 12, 14	I	Power ground
GNDS	1, 10 11, 20	I	Substrate ground
IN1	3	I	Control Input
IN2	6	I	Control Input
OUT1	5, 6	O	Half-H output. DMOS output
OUT2	15, 16	O	Half-H output. DMOS output
STATUS1	13	O	Status output
STATUS2	18	O	Latched status output
V$_{CC}$	2, 4, 12, 19	I	Supply voltage

NOTE: It is mandatory that all four ground terminals plus at least one substrate terminal are connected to the system ground. Use all V$_{CC}$ and OUT terminals.

FIGURE 5.26   Terminal functions of the TPIC0108B (image courtesy Texas Instruments).

TABLE 5.4   Inputs and Outputs for the TPIC0108B H-bridge

In1	In2	Out1	Out2	Mode
0	0	Z	Z	Quiescent supply current mode
0	1	LS	HS	Motor turns clockwise
1	0	HS	LS	Motor turns counterclockwise
1	1	HS	HS	Brake, both HSDs turned on hard

We just established a linear relationship between the back emf and the mechanical rotational velocity of the motor as

$$E_a = k_{emf}\omega_m \tag{5.21}$$

Here, we have k_{emf} as the back emf constant and ω_m as the motor shaft angular velocity. Combining the two equations and readjusting the terms gives us

$$\frac{di}{dt} = \frac{1}{L_a}[V - iR_a - k_{emf}\omega_m] \tag{5.22}$$

Assuming that there is no friction on the motor shaft, and the moment of inertia of the motor shaft and load combined is I_m, we get the following dynamic equation for the mechanical motion by applying Newton's law for rotation:

$$\frac{d\omega_m}{dt} = \frac{1}{I_m}\tau_m \tag{5.23}$$

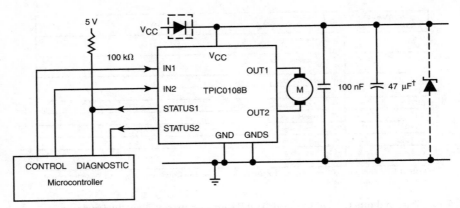

† Necessary for isolating supply voltage or interruption (e.g. 47 μF).

NOTE: If a STATUS output is not connected to the appropriate microcontroller input, it shall remain unconnected.

FIGURE 5.27 Sample application.

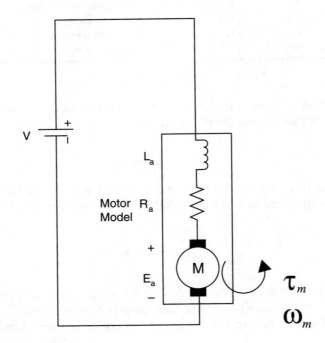

FIGURE 5.28 Basic motor circuit.

The application of (5.19) gives us

$$\frac{d\omega_m}{dt} = \frac{1}{I_m} k_{torque} i \tag{5.24}$$

Equations (5.22) and (5.24) give the electrical and mechanical dynamics of the system. This equation can be written in terms of the angle of the motor shaft as

$$\frac{d^2\theta_m}{dt^2} = \frac{1}{I_m} k_{torque} i \tag{5.25}$$

If we assume no loss of power, then all the electrical power used in the motor should convert to the mechanical power generated. The electrical power consumed is given by the product of the back emf and the current as

$$P_{electrical} = E_a i = k_{emf} \omega_m i \tag{5.26}$$

The dissipated mechanical power is given by

$$P_{mechanical} = \tau_m \omega_m = k_{torque} i \omega_m \tag{5.27}$$

Equating the electrical power consumed with the mechanical power generated, we get

$$k_{emf} = k_{torque} = k \tag{5.28}$$

We can make the model better by modeling friction terms in (5.24) that could include static and viscous friction terms:

$$\frac{d\omega_m}{dt} = \frac{1}{I_m} [ki - a \; \text{sign}(\omega_m) - b\omega_m] \tag{5.29}$$

In this equation, a is the static friction coefficient and b is the viscous friction coefficient. Now, let us study the dynamics when we add a gear that connects the motor shaft to a load. Let us use the angle of the gear on the motor side with a subscript m and the angle on the load side with subscript L. We will use the same subscripts for torques, angular velocities, and radii of the gears. Consider Fig. 5.29.

When the gear wheel on the motor side turns an angle θ_m, the distance traveled on its circumference is $r_m \theta_m$. The gear wheel on the load side that is

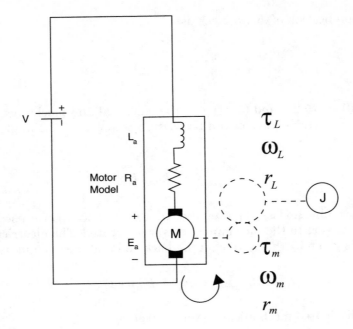

FIGURE 5.29 Gears.

connected to this gear should also have the same distance traveled on its circumference. Hence, we have

$$r_m \theta_m = r_L \theta_L \tag{5.30}$$

By differentiating both sides with respect to time, we get

$$r_m \omega_m = r_L \omega_L \tag{5.31}$$

We know that the force exerted by the first gear on the second has to be equal to the reaction force of the second gear on the first one. Hence, we should have

$$\frac{\tau_m}{r_m} = \frac{\tau_L}{r_L} \tag{5.32}$$

Multiplying (5.31) and (5.32) confirms the conservation of power principle:

$$\tau_m \omega_m = \tau_L \omega_L \tag{5.33}$$

Assuming a load with moment of inertia J and no friction terms, we obtain

$$J\frac{d\omega_L}{dt} = \tau_L \tag{5.34}$$

and

$$\tau_m = \left(\frac{r_m}{r_L}\right)\tau_L = \left(\frac{r_m}{r_L}\right)J\frac{d\omega_L}{dt} = \left(\frac{r_m}{r_L}\right)^2 J\frac{d\omega_m}{dt} \tag{5.35}$$

Using the torque constant, this can be written as

$$ki = R^2 J\frac{d\omega_m}{dt} \tag{5.36}$$

where R is the gear ratio given by

$$R = \frac{r_m}{r_L} \tag{5.37}$$

DC Motor Steady-State Analysis

Let us apply a constant voltage to a DC motor. Initially, the motor will accelerate and then reach a steady-state at some angular velocity. If we add a load torque to the shaft, the steady-state angular velocity will decrease. We can increase the load torque so much (for instance, by adding a big load on a rope) that the motor stops rotating. We can then plot the steady-state angular velocity with respect to the load torque. These are called the torque-speed characteristics. We can derive these using the dynamic equations. Assuming no static or viscous friction, when we apply the load torque T our dynamic equations become

$$\frac{di}{dt} = \frac{1}{L_a}[V - iR_a - k\omega_m] \tag{5.38}$$

and

$$\frac{d\omega_m}{dt} = \frac{1}{I_m}[ki - T] \tag{5.39}$$

In steady state, the rate of change of current and velocity becomes zero, and so we get

$$0 = \frac{1}{L_a}[V - iR_a - k\omega_m] \tag{5.40}$$

and

$$0 = \frac{1}{I_m}[ki - T] \tag{5.41}$$

Equation (5.41) gives us the torque in terms of the current. That expression can be substituted in (5.40) to give us

$$\omega_m = \frac{V}{k} - \frac{R_a}{k^2}T \tag{5.42}$$

This is a straight-line plot with a y-intercept of V/k and the x-intercept of $T = Vk/R_a$. This means that when there is no load torque (when the motor runs freely) by applying voltage V, the steady-state angular speed will be V/k. When we apply the minimum torque that makes the motor stop, the applied torque will be $T = Vk/R_a$. An example curve is shown in Fig. 5.30.

Identification of DC Motor Parameters

Various motor parameters can be obtained by manufacturer's sheets or experimental evaluation. These can be obtained as follows, experimentally.

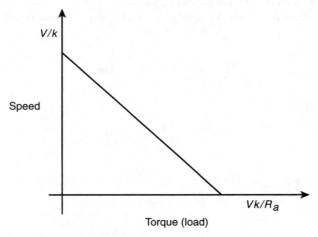

FIGURE 5.30 DC motor torque characteristic curve.

Motor Resistance

The motor resistance R_a can be obtained by simply using a multimeter to measure the terminal resistance across the motor lead.

Torque and Back Emf Constants

These constants are the same in *International System* (SI) units. The motor can be operated as a generator by connecting its shaft to another motor running at constant speed. Then, by measuring that speed and the voltage produced on the motor lead, we can calculate the motor constant as

$$k = \frac{E_a}{\omega_m} \tag{5.43}$$

Static Friction Coefficient

To obtain this coefficient, we study equation (5.29):

$$\frac{d\omega_m}{dt} = \frac{1}{I_m}[ki - a\,\text{sign}(\omega_m) - b\omega_m] \tag{5.44}$$

When we apply the minimum voltage, which barely starts the motor, the applied current is called the starting current. That current can be measured by a DC ammeter in series with the motor and the power applied (and the voltage by a DC voltmeter across the motor leads). In equation (5.44), this situation will have both sides at zero. Therefore, we will obtain

$$a = ki_{starting\ current} \tag{5.45}$$

Viscous Friction Coefficient

Once we know the torque constant and the static friction coefficient, we can obtain the viscous friction coefficient as well. Apply some voltage and when the system reaches a steady state measure the current as well as the steady-state angular velocity. Using (5.44) with the left side as zero, we get

$$b = \frac{1}{\omega_{steady-state}}[ki_{steady-state} - a] \tag{5.46}$$

Others

We can also obtain other parameters like the two time constants (the electric constant given by the ratio of the inductance and resistance, and the mechanical time constant given by the ratio of the viscous coefficient and the moment of inertia) by using advanced identification techniques using a frequency domain technique such as Bode plots, or by looking at transient analysis of the motor.

RC Servomotor

Radio controlled (RC) servomotors are designed to take a position command and then move the motor shaft to the commanded position. These servos are made from DC motors that have a potentiometer, which determines the angle of the motor shaft, and a servo controller chip that uses its associated peripheral components to take the commanded position in terms of a pulse and then drive the shaft to the commanded position in a feedback loop. The block diagram of an RC servomotor is shown in Fig. 5.31. A typical servo (made by Futaba) is shown in Fig. 5.32. A typical internal construction of a Futaba servo is shown in Fig. 5.33 and the components are given in Table 5.5.

The servo control is shown in Fig. 5.34. An RC servo has a limited motion, usually between −90 to +90 degrees. It has a mechanical stop to ensure that the motion is restricted. As seen in Figs. 5.33 and 5.34, a servo consists of a DC motor with some gearing connected to a potentiometer for feedback and some control circuitry. As an input, there are three wires. One wire is for power supply (usually 5V), another one for ground, and the third one for giving control signals. A control signal is a periodic pulse with a time period of approximately 20 ms. The position command is given by the width of the pulse. Usually a 1.5 ms pulse width indicates a 0-degree position (neutral) of the servo shaft. A pulse width of 1 ms indicates −90 degrees and a pulse width of 2 ms indicates +90 degrees. See Fig. 5.35.

The control command to the servo can come from a microprocessor/microcontroller, as in the case of the FLASH car, or some other circuit. The internal controller for the RC servo is a proportional controller that uses the poten-

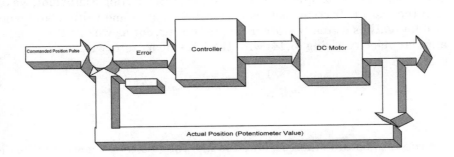

FIGURE 5.31   RC servo internal block diagram.

TABLE 5.5   The Components in the Futaba Servo

01	Upper case
02	Middle case
03	Bottom case
04	Metal bearing
05	Metal bearing
06	Potentiometer
07	Potentiometer drive plate
08	Motor
09	Motor pinion
10	Screw
11	1st gear
12	2nd gear
13	3rd gear
14	Final gear
15	Intermediate shaft
16	2nd shaft
17	Servo horn
18	Screw
19	Circuit board
20	Connector and cable
21	Cable bushing
22	Main assembly screw
23	Nameplate

FIGURE 5.32   Futaba servo.

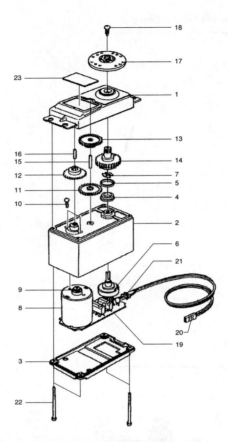

FIGURE 5.33   Futaba servo internals (image copyright © 2004 Futaba Corporation).

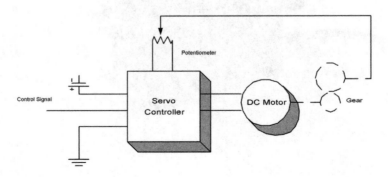

FIGURE 5.34   RC servo functional block diagram.

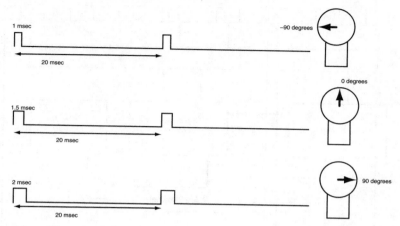

FIGURE 5.35   RC servo control signal.

tiometer feedback. Many servo controller chips can be used to design an RC servo. As an illustrative example, we will show how an *integrated circuit* (IC) controller NJM2611 chip can be used for this purpose. Our purpose is to show the internal function. As a designer of the FLASH car, we just need to produce the control input signals of Fig. 5.35. The internal block diagram of the RC servo controller chip is shown in Fig. 5.36.

The connection of this controller in a configuration of Fig. 5.35 is shown in Fig. 5.37. Here we can see the signal input in pin 1, the potentiometer connecting to pin 3, and also the internal and external transistors providing the H-bridge for the DC motor.

MATLAB Models

We will derive and use MATLAB models for the DC drive motor that we use in the FLASH car, as well as the Futaba servo we use for steering in the car.

MATLAB Model for DC Motor

We start by reproducing the dynamics of the DC motor. The electrical dynamics from equation (5.22) are given as

$$\frac{di}{dt} = \frac{1}{L_a}[V - iR_a - k_{EMF}\omega_m] \tag{5.47}$$

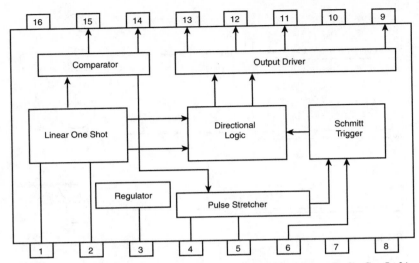

FIGURE 5.36 NJM2611 internal block diagram (image courtesy New Japan Radio Co., Ltd.).

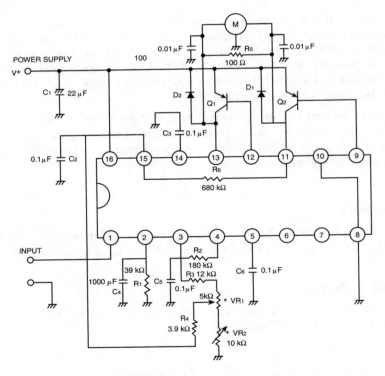

FIGURE 5.37 NJM2611 connection diagram (image courtesy New Japan Radio Co., Ltd.).

The mechanical dynamics from equation (5.29) are

$$\frac{d\omega_m}{dt} = \frac{1}{I_m}[ki - a\ \text{sign}(\omega_m) - b\omega_m] \qquad (5.48)$$

Taking a Laplace transform of (5.47) gives us

$$sL_aI(s) = V(s) - I(s)R_a - k_{emf}W_m(s) \qquad (5.49)$$

Here $I(s)$, $V(s)$, and $W(s)$ are the Laplace transforms of the time signals $i(t)$, $V(t)$, and $\omega(t)$, respectively. Rearranging terms in (5.49) gives us

$$[sL_a + R_a]I(s) = V(s) - k_{emf}W_m(s) \qquad (5.50)$$

After ignoring the static and viscous friction terms in equation (5.48), and taking its Laplace transform, it yields

$$s[sL_a + R_a]I_mW_m(s) = [V(s) - k_{emf}W_m(s)]k_{torque} \qquad (5.51)$$

By rearranging terms, we can obtain the transfer function from the applied voltage to the angular velocity as

$$G(s) = \frac{W_m(s)}{V(s)} = \frac{k_{torque}}{s[sL_a + R_a]I_m + k_{emf}k_{torque}} \qquad (5.52)$$

This simplifies to

$$G(s) = \frac{k_{torque}}{s^2L_aI_m + sR_aI_m + k_{emf}k_{torque}} \qquad (5.53)$$

which can be written as

$$G(s) = \frac{\dfrac{k_{torque}}{L_aI_m}}{s^2 + s\dfrac{R_a}{L_a} + \dfrac{k_{emf}k_{torque}}{L_aI_m}} \qquad (5.54)$$

The poles of the transfer functions are

$$p = \frac{1}{2}\left[-\frac{R_a}{L_a} \pm \sqrt{\left(\frac{R_a}{L_a}\right)^2 - 4\frac{k_{emf}k_{torque}}{L_aI_m}} \right] = \frac{1}{2}\frac{R_a}{L_a}\left[-1 \pm \sqrt{1 - 4\frac{k_{emf}k_{torque}L_a}{I_mR_a^2}} \right] \qquad (5.55)$$

Using the binomial expansion of the square root term, we can approximate this by

$$p \approx \frac{1}{2}\frac{R_a}{L_a}\left[-1 \pm \left(1 - 2\frac{k_{emf}k_{torque}L_a}{I_mR_a^2} \right) \right] \qquad (5.56)$$

Hence, the two poles are approximated as

$$p_{1,2} \approx -\frac{R_a}{L_a}, \frac{k_{emf}k_{torque}}{I_mR_a} \qquad (5.57)$$

The transfer function can be approximated as follows:

$$G(s) \approx \frac{\dfrac{k_{torque}}{L_aI_m}}{\left(s + \dfrac{R_a}{L_a} \right)\left(s + \dfrac{k_{emf}k_{torque}}{R_aI_m} \right)} \qquad (5.58)$$

This can also be written in terms of the electrical and mechanical time constants as

$$G(s) \approx \frac{\dfrac{1}{k_{emf}}}{(\tau_e s + 1)(\tau_m s + 1)} \qquad (5.59)$$

where the two time constants are

$$\tau_e = \frac{L_a}{R_a} \qquad (5.60)$$

$$\tau_m = \frac{R_aI_m}{k_{emf}k_{torque}} \qquad (5.61)$$

We use some typical parameters from F series Maxon DC motors. These are obtained from the data sheets of the motors (available on Maxon's Web site at www.maxonmotorusa.com/products/).

$$\tau_e = \frac{0.5mH}{9\Omega} = 0.55ms$$

$$\tau_m = 42ms$$

$$k_{emf} = 0.01Nm/A \tag{5.62}$$

These parameters give us the transfer function as

$$G(s) = \frac{4.33 \times 10^7}{(s + 18181.8)(s + 23.81)} \tag{5.63}$$

The MATLAB code for this transfer function is given here. The code inputs the transfer function and then calculates the Bode plot as well as the step response as shown in Figs. 5.38 and 5.39.

```
%DC Motor Model
num=[43300000]
den=[1,18205.61,432908.66]
system=tf(num,den)
poles=roots(den)
zpk(system)
```

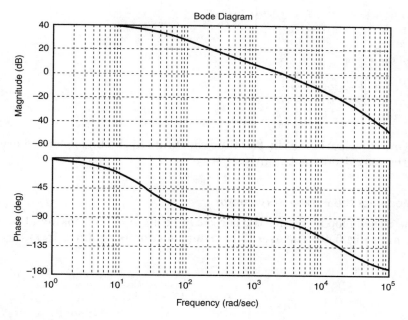

FIGURE 5.38   Bode plot for DC motor.

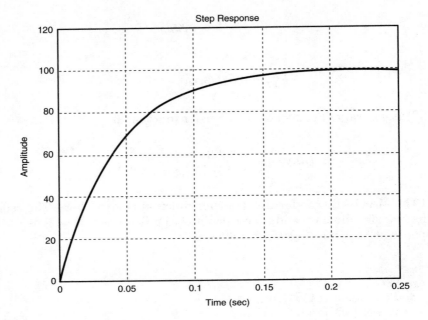

FIGURE 5.39   Step response for DC motor.

```
%[MAG,PHASE,W] = bode(system)
bode(system)
grid
pause

%Step Response
step(system)
grid
```

One can clearly see the two time constants in the Bode plots where the slope of the magnitude plot changes.

MATLAB Model for the Futaba Servomotor

The closed loop model for the servo can be obtained by looking at its block diagram in Fig. 5.40. The overall transfer function is obtained from

$$G(s) = \frac{\dfrac{k}{(\tau_e s + 1)(\tau_m s + 1)}}{1 + \dfrac{kk_f}{(\tau_e s + 1)(\tau_m s + 1)}} \tag{5.64}$$

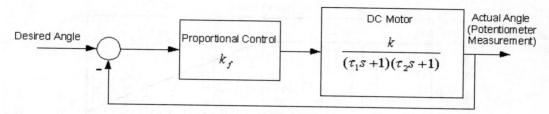

FIGURE 5.40 Servo block diagram.

This will be

$$G(s) = \frac{k}{(\tau_e s + 1)(\tau_m s + 1) + kk_f} \tag{5.65}$$

This is another second-order transfer function. We take the following estimated parameter model for the servo

$$G(s) = \frac{950}{s^2 + 40s + 950} \tag{5.66}$$

The following MATLAB code is the transfer function. The code inputs the transfer function and then calculates the Bode plot as well as the step response, as shown in Figs. 5.41 and 5.42:

```
%Servo Motor Model
num=[950]
den=[1,40,950]
system=tf(num,den)
poles=roots(den)
zpk(system)
%[MAG,PHASE,W] = bode(system)
bode(system)
grid
pause
%Step Response
step(system)
grid
```

Servo Hack

Many people find the way a microprocessor can send position commands to a servo to be convenient. Only a single port is required, and the signal is in terms of a PWM, as shown in Fig. 5.34. For this reason, and also for the fact that

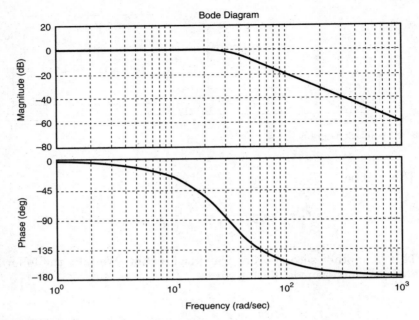

FIGURE 5.41   Bode plot for servomotor.

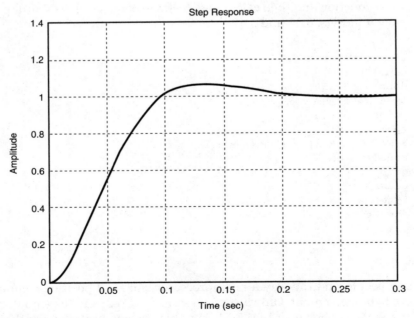

FIGURE 5.42   Step response for servomotor.

servos are, in general cheap and easily available, many people modify servos so that they can have continuous rotation instead of the rotation being limited to 180 degrees. They essentially modify a servo so that it can be used as a DC motor, but with one difference This motor can now be commanded to move forward and reverse by giving servo commands. In other words, the processor can give commands like the ones in Fig. 5.34, but the modified servo moves with constant forward speed or reverse, depending on the command. Essentially, the position commands have been converted into speed commands. In order to accomplish this, two modifications have to be performed on a servo:

- Servos can only move 180 degrees because the output gear shaft has a mechanical stop. In order to allow continuous rotation, we have to remove this tab. When you unscrew and open a Futaba servo, you can remove the gears and then physically remove the tab on the output shaft gear.

- The reason why a servo is able to follow a commanded angular position is that it measures the actual position using a potentiometer; then the difference between the commanded position and the reading of the potentiometer is used to make the servo shaft reach and maintain the correct position. The second modification we can do is to replace the potentiometer with fixed resistors as shown in Fig. 5.45. This way, the servo always thinks that the shaft is at the center position. So, when a processor gives it a positive angle command, the servo keeps rotating to make its measured position equal to the commanded one. However, since the reading is fixed, the error between the commanded position and the fixed reading is constant. Hence, the servo keeps getting commanded to move forward, and it keeps rotating forward. If you give a bigger forward command, it will move faster. If you give it a negative angle command, it will produce a continuous reverse rotation. Figs. 5.43 through 5.46 are pictures that show the internals of the Futaba servo and

FIGURE 5.43   The gear system inside the servo.

how to make the modifications. These are presented thanks to the courtesy of the Seattle Robotics Society and are also present at their Web site www.seattlerobotics.org.

Now that we know the physics behind DC motors and their controls, we can program microcontrollers to perform the motor actuation. The next chapter describes the PIC microcontroller and how it is used to control the FLASH car's DC motor and steering servo.

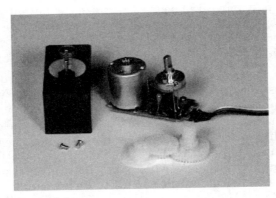

FIGURE 5.44 The servo disassembled.

FIGURE 5.45 Potentiometer replacement with 2.7KΩ resistors.

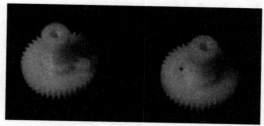

FIGURE 5.46 Gear tab removal, before (left) and after (right).

6

Low-Level Control

This chapter describes the PIC microcontroller program and how it interfaces with the rest of the car. As described in Chapter 2, "Overall System Structure," low-level control is done with the PIC16F874 microcontroller. The PIC acts as a converter between the *digital signal processor* (DSP) and the rest of the circuitry on the car. In this role, the PIC receives commands from the main processor, which is also called the DSP, and interprets those commands to generate signals that directly control the steering servo and motor. It also collects data from the optical encoder and sends it back to the DSP.

Tutorial

We begin with a brief tutorial on writing code for the PIC and downloading it to the chip. To go through this tutorial, you will need the following: a Microchip PIC16F874 microcontroller, a PC with the MPLAB *integrated development enviroment* (IDE) software (available for free from Microchip's Web site, www.microchip.com), and an MPLAB-compatible programmer (see the section on PIC programmers). The programmer requires a serial cable for connecting to the PC and a 15V DC power supply. To test the chip, you will need a 20 MHz crystal oscillator or frequency generator, a 5V DC power supply, and an oscilloscope.

As a preliminary step, power the PICSTART Plus with the 15V DC supply and connect it to the computer using the serial cable.

The steps required in the process of generating code are shown in Fig. 6.1. First, the assembly code must be written and saved as an .asm file. Then, in MPLAB, the code must be assembled into a file format that the PIC itself can understand, known as a HEX file. Finally, the code must be downloaded from the PC to the PIC using the programmer.

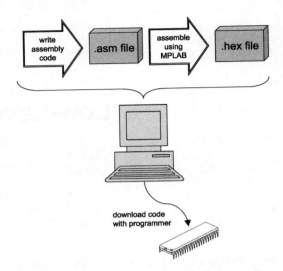

FIGURE 6.1 The steps in the process of programming the PIC microcontroller.

First, we will go through the process of writing the code in MPLAB. The code is simply text and can be written using any editor. If you choose to use another text editor, make sure you save the file with an .asm extension so that it will be recognized by MPLAB. Open MPLAB and start a new project by going to the Project menu and choosing New. Give the project a name and select the directory where you want to save it. Then go to Configure/Select Device and choose the PIC16F874. This selects the particular PIC that we are using. Click OK.

The next step is to type in the code. Go to File/New to bring up a text editor window. In this editor, type the following code. Make sure that the lines that are flush left have no spaces in front of them and that the other lines have at least one space in front of them.

```
     list p=16f874      ; set processor type
     include <P16f874.INC>

count equ 0x20

     org 00000h         ; Reset Vector
     goto Start

     org 00004h         ; Interrupt vector
     goto IntVector

; ***** MAIN PROGRAM *****
     org 00020h ;Beginning of program EPROM
```

```
    Start
        bsf STATUS, RP0      ; Switch to Bank 1
        movlw b'00000000'
        movwf OPTION_ REG    ; Configure TMR0
        movlw b'11111100'
        movwf TRISB ; Set bits 0 and 1 of PORTB to be outputs
        bcf STATUS, RP0      ; Switch to Bank 0

        movlw b'10100000'
        movwf INTCON         ; Enable TMR0 interrupt
        movlw d'64'
        movwf count          ; Initialize count to 64
        movwf TMR0           ; Load count value into TMR0

        bsf PORTB,1          ; Set PORTB pin 1

    Main
        bsf PORTB,0          ; Set PORTB pin 0
        nop
        nop
        nop
        nop
        bcf PORTB,0          ; Clear PORTB pin 0
        goto Main

    IntVector
        bcf INTCON,2         ; Clear the interrupt
        btfss PORTB,1        ; Check value of PORTB pin 1
        goto LOADHIGH
        goto LOADLOW

    LOADHIGH
        bsf PORTB,1          ; Set PORTB pin 1
        movf count,0
        movwf TMR0           ; Load count into TMR0
        goto RET

    LOADLOW
        bcf PORTB,1          ; Clear PORTB pin 1
        movf count,0
        sublw 0
        movwf TMR0           ; Load -count into TMR0

    RET
        retfie               ; Return from interrupt

        end
```

If you have never programmed in assembly before, this code may look like gibberish. Don't worry; the syntax and instruction set of the PIC assembly language are described in more detail later in this chapter. For now, we are just going through the process of using MPLAB to program the PIC.

Now that the code is written, go to File/Save As and give the file a name with an .asm extension. Make sure you save it to the correct directory. Next, go to Project/Add Files to Project and then select the .asm file you just created. Click Open. The project is now ready to be assembled into a HEX file. Go to Project/Build All. This opens the Output window. If you typed the code correctly, the window should show that the build was completed successfully. If there were errors in the code, the build results indicate on which line the errors occurred.

Once the code is assembled successfully, go to the Programmer menu and select the PICSTART Plus. Then go to Programmer/Enable Programmer. If communication is not established, check which serial port the programmer is plugged into and make sure MPLAB is using that port. When the programmer is enabled, insert the PIC into it and go to Programmer/Program to download the code. When the download is complete, the Output window indicates whether the chip was successfully programmed. If it was not successful, make sure the chip was inserted into the programmer with the proper orientation.

To test the code, set up the circuit as shown in Fig. 6.2. When the circuit is powered, on Channel 1 you should see a 625 kHz signal that has a high time of about 1 μs. On Channel 2 you should see a 9.3 kHz signal with a high time of about 80 μs.

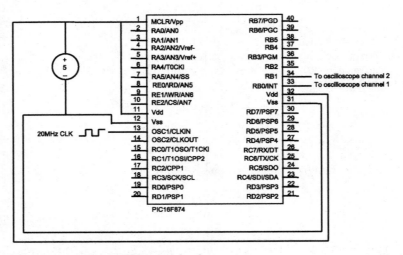

FIGURE 6.2   The test circuit for the PIC tutorial code.

Features of the PIC16F874

The PIC must perform several tasks simultaneously. It must communicate with the DSP so that it knows how to steer and how fast to turn the motor. It must interpret commands from the DSP into pulse widths. It must generate PWM signals for the motor and servo. It must count the number of pulses coming from the optical encoder to determine the speed of the rear wheels. It must send the wheel speed back to the DSP. How does a little chip do so many things so seamlessly? The answer lies in three features of the PIC: the interrupts, the timers, and the *parallel slave port* (PSP).

Interrupts.   Several things can cause interrupts on the PIC; for example, a particular pin value may change, or a certain register may overflow. The former is known as an external interrupt because it is generated by a source outside the PIC. The latter is called an internal interrupt because it is generated within the PIC itself.

When an interrupt is generated, the PIC stops what it is doing, goes to a particular memory location, and executes the code that is at that location. This code is called the *interrupt service routine* (ISR). When the ISR is complete, the PIC returns to what it was doing when the interrupt was generated.

There are two external interrupts on this PIC. One is the dedicated external interrupt (pin 33). This can be configured to cause an interrupt on either the rising or falling edge. The other external interrupt is RB port change. When enabled, an interrupt occurs when at least one of the PORTB bits 4, 5, 6, or 7 has changed.

Timers.   This particular model PIC has three built-in timer modules: TMR0, TMR1, and TMR2. Each timer is similar in operation but has its own unique features that are utilized on the car.

TMR0 and TMR1 are similar, except TMR0 is an 8-bit register while TMR1 consists of two 8-bit registers. Both can be configured as timers or counters. When configured as a timer, the register increments every instruction cycle. With the modification of its associated parameter, which is known as a prescaler, the register can be made to increment every 2, 4, 8, 16, 32, 64, 128, or 256 instruction cycles. When configured as counters, TMR0 increments on the rising (or falling) edge of T0CKI (pin 6, see Fig. 6.3) and TMR1 increments on the rising (or falling) edge of T1CKI (pin 15).

There are also interrupts associated with both TMR0 and TMR1. When enabled, an interrupt is generated when the TMR register overflows. For TMR0, the interrupt is generated when the register goes from 255 to 0. For TMR1, the interrupt is generated when the register goes from 65,535 to 0. (This is because TMR0 has 8 bits and TMR1 has 16.)

```
  1    MCLR/Vpp              RB7/PGD    40
  2    RA0/AN0               RB6/PGC    39
  3    RA1/AN1                  RB5     38
  4    RA2/AN2/Vref-            RB4     37
  5    RA3/AN3/Vref+          RB3/PGM   36
  6    RA4/T0CKI                RB2     35
  7    RA5/AN4/SS               RB1     34
  8    RE0/\RD/AN5            RB0/INT   33
  9    RE1/\WR/AN6              Vdd     32
 10    RE2/\CS/AN7              Vss     31
 11    Vdd                   RD7/PSP7   30
 12    Vss                   RD6/PSP6   29
 13    OSC1/CLKIN            RD5/PSP5   28
 14    OSC2/CLKOUT           RD4/PSP4   27
 15    RC0/T1OSO/T1CKI       RC7/RX/DT  26
 16    RC1/T1OSI/CPP2        RC6/TX/CK  25
 17    RC2/CPP1              RC5/SDO    24
 18    RC3/SCK/SCL           RC4/SDI/SDA 23
 19    RD0/PSP0              RD3/PSP3   22
 20    RD1/PSP1              RD2/PSP2   21
```

PIC16F874

FIGURE 6.3   The pinout for the PIC16F874.

On the FLASH car, TMR0 is configured as a timer and is used to generate the servo control pulse. TMR1 is configured as a counter and counts the pulses from the optical encoder to measure the wheel speed.

TMR2 is the PWM module. This timer is very useful on the car because it generates a PWM signal in the background with minimal programming effort. Once the TMR2 module is configured as for PWM, it is only necessary to write a 10-bit value to a holding register that represents the duty cycle. The module toggles CCP1 (pin 17) automatically to generate the PWM signal. After one period of the PWM signal, the duty cycle is read from the holding register. This scheme allows the desired duty cycle to be loaded into the register at any time and the PWM output will be modified on the very next PWM period. TMR2 is used to generate the PWM signal that drives the motor on the car.

PSP.   The PIC has five input/output ports, labeled PORTA, PORTB, PORTC, PORTD, and PORTE. As with the timers, these ports share some common features while each has its own unique and useful characteristics. For example, they can all have their individual pins assigned as either inputs or outputs; for example, PORTA bits 1, 2, and 4 can be configured as inputs while pins 0, 3, and 5 can be configured as outputs.

Communication with the DSP is done with the PSP, PORTD. When configured as the PSP, PORTD can be connected directly to the data bus and controlled using $\overline{CS}$, $\overline{RD}$, and $\overline{WR}$. The value that is to be sent from the PIC to the DSP is loaded into the PORTD output buffer. When the $\overline{CS}$ and $\overline{RD}$ pins both go low, the values in the PORTD output buffer are transferred to the PORTD pins (pins 19 through 22 and 28 through 30) and are available on the data bus.

When the DSP must send a value to the PIC, the DSP makes the $\overline{CS}$ and $\overline{WR}$ pins low, puts the value on the data bus, and then makes the $\overline{CS}$ and $\overline{WR}$ pins high. On the PIC, when the $\overline{CS}$ and $\overline{WR}$ pins are low, the values on the PORTD pins are transferred to the PORTD input buffer. When the $\overline{CS}$ and $\overline{WR}$ go high, this indicates that the write is complete and an interrupt is generated. Then, in the interrupt service routine for this particular interrupt, the value that was written can be read from the input buffer into a variable. The input and output buffers are completely separate inside the PIC so that there are no collisions when reading and writing data.

Program Details

Now that we have some idea of how the PIC performs all its required tasks, how do the pieces fit together? In this section we go through and describe the code contained in each subroutine. The PIC is programmed in assembly using its 35 instructions, which can be rather tedious. While going through the various subroutines, the basics of the PIC's registers and syntax are described as appropriate. However, this section is not intended to be a complete manual on how to program the PIC. For a complete description of the PIC's registers and instruction set, see the PIC16F874 datasheet. In addition, there is an in-depth guide to PIC programming called *Programming and Customizing the PIC Microcontroller* (Predko 1998). This book goes into detail about the inner workings of the PIC and provides several projects and experiments designed to teach how the PIC works.

Initialization

Initialization begins with assembler directives.

Assembler directives. At the very beginning of the code are the assembler directives. These control how the code is assembled, but are not part of the actual code that is executed on the PIC. The assembler is described in the section on programming the PIC. The program starts with the following two lines:

```
list p=16f874 ; set processor type
include <P16f874.INC>
```

The first line tells the assembler for which particular PIC device the code is intended. The second line causes the file "P16f874.INC" to be inserted at that point in the code as if it were typed in there. This "include" command works the same way as "#include" used in C or C++.

The P16f874.INC file contains assignments so that particular registers and bits can be called by their names rather than their memory addresses. For example, the carry bit of the STATUS register can be referred to as STATUS,C rather than H'0003',0. Here H'0003' is the memory location of the STATUS register and the carry bit is bit 0.

Next the program's variables are declared using the "equ" directive:

```
highspeed     equ 0020h
steer         equ 0021h
steercnt      equ 0022h
W_ TEMP       equ 0023h
STATUS_ TEMP  equ 0024h
hold          equ 0025h
loop_ 0       equ 0026h
cmdword       equ 0027h
temp1         equ 0028h
r_ speed      equ 0029h
```

This assigns names to memory addresses. So when a value is written to *highspeed*, it is stored in memory location 0020h. One note about notation: H'0020', 0020h, and 0×0020 are interpreted the same. All indicate that the number given is represented in hexadecimal format.

Also just as in C, a "#define" can be used to simplify the code by using the text substitution:

```
#DEFINE MOTORDIR PORTC,1
#DEFINE SERVOPIN PORTC,4
#DEFINE HEART    PORTC,3
```

Whenever MOTORDIR is encountered in the code, it is replaced by PORTC,1. This helps to make the code more readable. In addition, if one decided to reassign the motor pin, the only code that must be changed is in the #define statement. This is much easier than searching through the code to find and change every mention of PORTC,1.

Next comes the declaration of the reset and interrupt vectors:

```
org   00000h      ; Reset Vector
goto Start

org   00004h      ; Interrupt vector
goto IntVector
```

When the power is turned on or the PIC is reset, the program counter goes to memory location 0000h. Then the "goto Start" command is executed, which sends the program counter to the beginning of the main program. When an interrupt occurs, the program counter is loaded with 0004h. In this case, the "goto IntVector" command is executed, which sends the program counter to the interrupt service routine. In other words, "org" places code at a certain memory location. When the program counter goes to that memory location, that particular line of code is executed.

It is also important to notice that a semicolon indicates a comment. Everything that follows on the same line as a semicolon is ignored by the assembler.

Variable initialization. All of the above code (except for the goto statements) is not actually PIC instructions. They are assembler directives that tell how the actual PIC code should be arranged on the chip. The actual PIC code is next and starts with the following:

```
    org 00020h ; Beginning of program EPROM
Start
;Initialize Variables
    movlw   0x00
    movwf   highspeed
    movlw   0x02
    movwf   hold
    movlw   d'78'
    movwf   steercnt
    movlw   d'7'
    movwf   loop_ 0
    bcf     MOTORDIR
    bcf     SERVOPIN
```

Again, at the top of this section of code is the "org" command. That tells the assembler that the first line of code following that directive, "movlw 0×00" should be placed at memory location 0020h.

"Start" is a label that gives a name to a piece of code. Labels are for ease of programming so that you can write "goto Start" rather than "goto 0020h." This is convenient since the actual memory location of a labeled instruction may move depending on how much code is placed before it. It is required that labels start in the first column of text. After the label is a commented line indicating that what follows is variable initialization.

Then the actual PIC commands begin. It is required that the instructions begin in the second column or later. Some instructions have operands. There must be at least one space between the instruction and the operand.

On the PIC, you cannot load values directly into a register. Rather, you can only load values in the working register (denoted by "W," also known as the accumulator). This makes initializing the variables a little bit tedious. For example, the first two lines of code move the value 0 into the variable *highspeed*. The first line moves the literal "0" into the working register using the *move literal to W* (movlw) command. The second line moves the working register value into the register file denoted by *highspeed* using the *move W to register file* (movwf) command. In the first case, the operand is the literal that is to be moved into W. In the second case, the operand is the register file address where the value is to be moved.

The six lines of code that follow this initialize the variables *hold, steercnt*, and *loop_0* to 2, 78, and 7, respectively.

Finally, individual bits of a register can be set or cleared using the *bit set register file* (bsf) and *bit clear register file* (bcf) commands. In the previous case, PORTC bits 1 and 4 (denoted by MOTORDIR and SERVOPIN) are cleared, or, made to be logic 0.

TMR0. TMR0 is used to generate the servo control pulse. To do this, TMR0 is configured as a timer with its interrupt enabled and its prescaler set to a value so that the servo control pulse period is about 10 ms. The following code sets up TMR0:

```
; Set up tmr0 for SCP at 84.75Hz
    bsf     STATUS,RP0
    movlw   0xd4             ; Set TMR0 for prescaler=32
    movwf   OPTION_ REG
    movlw   0xa0             ; Enable global and TMR0 interrupt
    movwf   INTCON
    bcf     STATUS,RP0
```

There are four memory banks on this PIC: banks 0 to 3. The address that you use as an instruction's operand is appended to bits RP0 and RP1 of the STATUS register. When writing to or reading from a register, you must make sure that you are in the correct memory bank. Otherwise, you may inadvertently be accessing another register. In this program, all the registers are in bank 0 or 1. To access bank 0, the RP0 bit of the STATUS register must be made a 0 by using the bcf command. To access bank 1, RP0 must be a 1.

The OPTION_REG register contains the configuration bits for TMR0 and the external interrupt. The bits of this register are shown in Fig. 6.4. The INTCON register is used to configure the bits related to the TMR0 and external interrupts and is shown in Fig. 6.5. Both the OPTION_REG and INTCON registers are located in bank 1, so first RP0 must be set to a 1. Then the value 11010100 is loaded into the OPTION_REG. This makes TMR0 increment with the instruction clock (as opposed to an external clock) and sets the prescaler value to 32. The value 10100000 is loaded into the INTCON register, enabling global interrupts and, specifically, the TMR0 interrupt.

TMR1. TMR1 is a 16-bit timer that is used to count the forward pulses of the optical encoder. The following code initializes the timer for this use:

```
; Set up timer 1 to count encoder "Up" pulses
     clrf      TMR1L
     clrf      TMR1H
     movlw     0x07
     movwf     T1CON
```

REGISTER 2-2:   OPTION_REG REGISTER (ADDRESS 81h, 181h)

R/W-1	R/W-1	R/W-1	R/W-1	R/W-1	R/W-1	R/W-1	R/W-1
RBPU	INTEDG	T0CS	T0SE	PSA	PS2	PS1	PS0

bit 7   bit 0

bit 7   **RBPU:** PORTB Pull-up Enable bit
         1 = PORTB pull-ups are disabled
         0 = PORTB pull-ups are enabled by individual port latch values

bit 6   **INTEDG:** Interrupt Edge Select bit
         1 = Interrupt on rising edge of RB0/INT pin
         0 = Interrupt on falling edge of RB0/INT pin

bit 5   **T0CS:** TMR0 Clock Source Select bit
         1 = Transition on RA4/T0CKI pin
         0 = Internal instruction cycle clock (CLKOUT)

bit 4   **T0SE:** TMR0 Source Edge Select bit
         1 = Increment on high-to-low transition on RA4/T0CKI pin
         0 = Increment on low-to-high transition on RA4/T0CKI pin

bit 3   **PSA:** Prescaler Assignment bit
         1 = Prescaler is assigned to the WDT
         0 = Prescaler is assigned to the Timer0 module

bit 2-0   **PS2:PS0:** Prescaler Rate Select bits

Bit Value	TMR0 Rate	WDT Rate
000	1 : 2	1 : 1
001	1 : 4	1 : 2
010	1 : 8	1 : 4
011	1 : 16	1 : 8
100	1 : 32	1 : 16
101	1 : 64	1 : 32
110	1 : 128	1 : 64
111	1 : 256	1 : 128

Legend:
R = Readable bit   W = Writable bit   U = Unimplemented bit, read as '0'
- n = Value at POR   '1' = Bit is set   '0' = Bit is cleared   x = Bit is unknown

FIGURE 6.4 The bits in the OPTION_REG register (image courtesy Microchip Technology Inc.).

REGISTER 2-3:   **INTCON REGISTER (ADDRESS 0Bh, 8Bh, 10Bh, 18Bh)**

R/W-0	R/W-0	R/W-0	R/W-0	R/W-0	R/W-0	R/W-0	R/W-x
GIE	PEIE	T0IE	INTE	RBIE	T0IF	INTF	RBIF

bit 7   bit 0

bit 7   **GIE:** Global Interrupt Enable bit
                1 = Enables all unmasked interrupts
                0 = Disables all interrupts

bit 6   **PEIE:** Peripheral Interrupt Enable bit
                1 = Enables all unmasked peripheral interrupts
                0 = Disables all peripheral interrupts

bit 5   **T0IE:** TMR0 Overflow Interrupt Enable bit
                1 = Enables the TMR0 interrupt
                0 = Disables the TMR0 interrupt

bit 4   **INTE:** RB0/INT External Interrupt Enable bit
                1 = Enables the RB0/INT external interrupt
                0 = Disables the RB0/INT external interrupt

bit 3   **RBIE:** RB Port Change Interrupt Enable bit
                1 = Enables the RB port change interrupt
                0 = Disables the RB port change interrupt

bit 2   **T0IF:** TMR0 Overflow Interrupt Flag bit
                1 = TMR0 register has overflowed (must be cleared in software)
                0 = TMR0 register did not overflow

bit 1   **INTF:** RB0/INT External Interrupt Flag bit
                1 = The RB0/INT external interrupt occurred (must be cleared in software)
                0 = The RB0/INT external interrupt did not occur

bit 0   **RBIF:** RB Port Change Interrupt Flag bit
                1 = At least one of the RB7:RB4 pins changed state; a mismatch condition will continue to set
                    the bit. Reading PORTB will end the mismatch condition and allow the bit to be cleared
                    (must be cleared in software).
                0 = None of the RB7:RB4 pins have changed state

FIGURE 6.5 The bits in the INTCON register (image courtesy Microchip Technology Inc.).

Since the PIC is an 8-bit device, TMR1 consists of two 8-bit registers, TMR1L and TMR1H. The first two lines of code clear these registers using the *clear register file* (clrf) command. Next, the value 00000111 is loaded into the T1CON register. This register is shown in Fig. 6.6. This value enables the timer, sets the prescaler to 1, and chooses the external clock as its counting source. When the rear wheels turn forward, the TMR1L increments on the rising edge.

TMR2.   TMR2 can be used as a dedicated PWM generator. The following code configures TMR2 to output a PWM signal with a frequency of 1.22 kHz:

```
; Set up TMR2 for use as the PWM generator at 1.22 kHz
; for the H-bridge, and set to 0
   bsf     STATUS,RP0
   movlw   0xFF
   movwf   PR2
   bcf     STATUS,RP0
   movlw   0x07
   movwf   T2CON
   movlw   0x0C
```

REGISTER 6-1:   **T1CON: TIMER1 CONTROL REGISTER (ADDRESS 10h)**

U-0	U-0	R/W-0	R/W-0	R/W-0	R/W-0	R/W-0	R/W-0
—	—	T1CKPS1	T1CKPS0	T1OSCEN	T1SYNC	TMR1CS	TMR1ON

bit 7   bit 0

bit 7-6	Unimplemented: Read as '0'
bit 5-4	T1CKPS1:T1CKPS0: Timer1 Input Clock Prescale Select bits
	11 = 1:8 Prescale value
	10 = 1:4 Prescale value
	01 = 1:2 Prescale value
	00 = 1:1 Prescale value
bit 3	T1OSCEN: Timer1 Oscillator Enable Control bit
	1 = Oscillator is enabled
	0 = Oscillator is shut-off (the oscillator inverter is turned off to eliminate power drain)
bit 2	**T1SYNC:** Timer1 External Clock Input Synchronization Control bit
	When TMR1CS = 1:
	1 = Do not synchronize external clock input
	0 = Synchronize external clock input
	When TMR1CS=0:
	This bit is ignored. Timer1 uses the internal clock when TMR1CS=0.
bit 1	TMR1CS: Timer1 Clock Source Select bit
	1 = External clock from pin RC0/T1OSO/T1CKI (on the rising edge)
	0 = Internal clock (FOSC/4)
bit 0	TMR1ON: Timer1 On bit
	1 = Enables Timer1
	0 = Stops Timer1

FIGURE 6.6 The bits in the T1CON register (image courtesy Microchip Technology Inc.).

```
movwf   CCP1CON
movlw   0x00
movwf   CCPR1L
```

First, the value $0 \times FF$ is loaded into the PR2 register, which is located in bank 1. Then the value 00000111 is loaded into the T2CON register. This enables TMR2 and makes the prescaler 16. The PWM period is given by

$$T = (PR2 + 1)*4*T_{osc}*TMR2_{precaler} \tag{6.1}$$

where T_{osc} is $0.05\,\mu s$ for a 20 MHz clock.

I/O Ports.   The use of each of the I/O ports is given as follows:

PortA :   Not in use

PortB :   0 - Used as an interrupt to measure reverse speed

                1...5 - Not in use

PortC :   0 - Encoder

                1 - Motor Direction

                2 - Motor Speed

                3 - Heartbeat

                4 - Servo Control

                5...7 - Not in use

PortD :   0...7 - Speed and steering input from and real speed output to PSP
PortE :   0 - PSP Read Pin
           1 - PSP Write Pin
           2 - PSP Enable

By default, all of the I/O ports are initialized as inputs. PORTA is not used, and so does not need to be configured. On PORTB, only bit 0 is used as the external interrupt. PORTC contains the signals that interface with the motor, servo, and encoder. PORTD and PORTE together make up the PSP and must be configured as such. The following sections detail the initialization of these I/O ports.

RB0.   RB0 is used as the external interrupt to count the reverse pulses from the encoder. The reverse encoder count is stored in the variable *r_speed*. The following code does the initialization:

```
; Set up RB0/Int to count encoder "Up" pulses in reverse
    clrf    r_ speed
    bsf     STATUS,RP0
    bsf     OPTION_ REG,6
    bsf     INTCON,4
    bcf     STATUS,RP0
```

First, *r_speed* is set to zero. Then, by setting bit 6 of the OPTION_REG register, the interrupt is configured to occur on the rising edge of the signal. Then, setting bit 4 of the INTCON register enables the external interrupt. See Fig. 6.4 and Fig. 6.5. It does not matter in which order these bits are set. Notice again that these registers are in bank 1, so the RP0 bit must be changed accordingly. Also notice that configurations for the external interrupt and TMR0 can be done simultaneously by writing the appropriate values to the OPTION_REG and INTCON registers once. It may be better to write these values at the same time, because if you do them separately, you may inadvertently undo some configuration. However, the initialization is divided into these to pieces to make their explanation clearer.

PORTC.   For PORTC, bit 0 is an input from the encoder while bits 1 through 4 are outputs. So bits 1 through 4 must be changed from their initial state as inputs to outputs. This is done as follows:

```
; Set up PORTC,0 as input for encoder pulses
    bsf     STATUS,RP0
    clrf    TRISC
    bsf     TRISC,0
    bcf     STATUS,RP0
```

TRISC is the data direction register for PORTC. It is located in memory bank 1. If a bit in TRISC is a zero, then the corresponding bit of PORTC is an output. If a bit in TRISC is a one, then the corresponding bit of PORTC is an input. Initially, TRISC is all ones. So first the RP0 bit is made a 1. Then TRISC is made all zeros, thus making all of PORTC outputs. Bit 0 is then changed back to an input by setting the bit. This is one of many ways to configure the port. For example, the same PORTC configuration would result from simply writing the value 00000001 to the working register and then moving it to TRISC. It is really a matter of style and preference in choosing which specific commands to use.

PSP. PORTD and PORTE together comprise the PSP. PORTD contains the eight bits over which the data is transferred to and from the DSP. Bits 0, 1, and 2 of PORTE are respectively the $\overline{RD}$, $\overline{WR}$, and $\overline{CS}$ control bits for the PSP. The following code configures the PSP:

```
; Set up the PSP to allow DSP to communicate
; with the PIC.
    bsf STATUS,RP0
    movlw 0x17          ; Enable the PSP and configure
    movwf TRISE         ; PORTE as inputs
    movlw 0x07
    movwf ADCON1        ; Configure RE0, RE1, RE2 as digital I/O
    bcf STATUS,RP0
```

PSP mode is enabled by setting bit 4 of the TRISE register. Then PORTE bits 0, 1, and 2 must be made inputs by setting the corresponding bits in TRISE. See Fig. 6.7. This is done by simply loading the value 00010111 into the TRISE register. Contrast this with the configuration of TRISC earlier.

Then, the value 00000111 is loaded into the ADCON1 register. This register configures the on-chip 10-bit *analog-to-digital* (A/D) converter. Bits 0, 1, and 2 of PORTE are three of the analog inputs for the A/D. However, we are using an external A/D converter on the car and require the bits of PORTE to be the $\overline{RD}$, $\overline{WR}$, and $\overline{CS}$ control bits for the PSP.

Interrupt Enabling. The final step in the program initialization is the interrupt configuration:

```
; Clearing Interrupt Flags, and setting Interrupts
    movlw 0x00
    movwf PIR1          ; Clears all interrupt flags
    bsf STATUS,RP0
    movlw 0x81
    movwf PIE1          ; Sets PSP and TMR1 interrupt enables
```

REGISTER 3-1:   TRISE REGISTER (ADDRESS 89h)

R-0	R-0	R/W-0	R/W-0	U-0	R/W-1	R/W-1	R/W-1
IBF	OBF	IBOV	PSPMODE	6	Bit2	Bit1	Bit0

bit 7   bit 0

Parallel Slave Port Status/Control Bits:

bit 7   **IBF:** Input Buffer Full Status bit
1 = A word has been received and is waiting to be read by the CPU
0 = No word has been received

bit 6   **OBF:** Output Buffer Full Status bit
1 = The output buffer still holds a previously written word
0 = The output buffer has been read

bit 5   **IBOV:** Input Buffer Overflow Detect bit (in Microprocessor mode)
1 = A write occurred when a previously input word has not been read (must be cleared in software)
0 = No overflow occurred

bit 4   **PSPMODE:** Parallel Slave Port Mode Select bit
1 = PORTD functions in Parallel Slave Port mode
0 = PORTD functions in general purpose I/O mode

bit 3   **Unimplemented:** Read as '0'

PORTE Data Direction Bits:

bit 2   **Bit2:** Direction Control bit for pin RE2/$\overline{CS}$/AN7
1 = Input
0 = Output

bit 1   **Bit1:** Direction Control bit for pin RE1/$\overline{WR}$/AN6
1 = Input
0 = Output

bit 0   **Bit0:** Direction Control bit for pin RE0/$\overline{RD}$/AN5
1 = Input
0 = Output

FIGURE 6.7   The bits in the TRISE register (image courtesy Microchip Technology Inc.).

```
bcf STATUS,RP0

bsf INTCON,PEIE ; Enable peripheral interrupts
bsf INTCON,GIE  ; Enable global interrupts
```

The PIR1 register contains all the flag bits for the peripheral interrupts. They are set to be all zeros so that the program doesn't jump to the interrupt service routine inadvertently. The PIE1 is used to enable each peripheral interrupt individually. See Figs. 6.8 and 6.9. Notice that the bits for the RB0 and TMR0 interrupts are located separately in the INTCON register. When enabling individual interrupts, it is important to set the appropriate bits in the INTCON and PIE1 registers. Also, the global interrupt enable bit (INTCON, bit 7) must be set for any of the interrupts to work and the peripheral interrupt enable bit (INTCON, bit 6) must be set for any of the peripheral interrupts to work.

Main Program Loop

The flowchart in Fig. 6.10 shows the main program of the PIC. As you can see, not much happens in the main program: PORTC bit 3 is toggled each time through the loop. This is done for debugging purposes. When the car is turned

REGISTER 2-5:   **PIR1 REGISTER (ADDRESS 0Ch)**

R/W-0	R/W-0	R-0	R-0	R/W-0	R/W-0	R/W-0	R/W-0
PSPIF[1]	ADIF	RCIF	TXIF	SSPIF	CCP1IF	TMR2IF	TMR1IF

bit 7   bit 0

bit 7   **PSPIF[1]**: Parallel Slave Port Read/Write Interrupt Flag bit
1 = A read or a write operation has taken place (must be cleared in software)
0 = No read or write has occurred

bit 6   **ADIF**: A/D Converter Interrupt Flag bit
1 = An A/D conversion completed
0 = The A/D conversion is not complete

bit 5   **RCIF**: USART Receive Interrupt Flag bit
1 = The USART receive buffer is full
0 = The USART receive buffer is empty

bit 4   **TXIF**: USART Transmit Interrupt Flag bit
1 = The USART transmit buffer is empty
0 = The USART transmit buffer is full

bit 3   **SSPIF**: Synchronous Serial Port (SSP) Interrupt Flag
1 = The SSP interrupt condition has occurred, and must be cleared in software before returning
from the Interrupt Service Routine. The conditions that will set this bit are:
  □ SPI
     - A transmission/reception has taken place.
  □ I^{2}C Slave
     - A transmission/reception has taken place.
  □ I^{2}C Master
     - A transmission/reception has taken place.
     - The initiated START condition was completed by the SSP module.
     - The initiated STOP condition was completed by the SSP module.
     - The initiated Restart condition was completed by the SSP module.
     - The initiated Acknowledge condition was completed by the SSP module.
     - A START condition occurred while the SSP module was idle (Multi-Master system).
     - A STOP condition occurred while the SSP module was idle (Multi-Master system).
0 = No SSP interrupt condition has occurred.

bit 2   **CCP1IF**: CCP1 Interrupt Flag bit
<u>Capture mode:</u>
1 = A TMR1 register capture occurred (must be cleared in software)
0 = No TMR1 register capture occurred
<u>Compare mode:</u>
1 = A TMR1 register compare match occurred (must be cleared in software)
0 = No TMR1 register compare match occurred
<u>PWM mode:</u>
Unused in this mode

bit 1   **TMR2IF**: TMR2 to PR2 Match Interrupt Flag bit
1 = TMR2 to PR2 match occurred (must be cleared in software)
0 = No TMR2 to PR2 match occurred

bit 0   **TMR1IF**: TMR1 Overflow Interrupt Flag bit
1 = TMR1 register overflowed (must be cleared in software)
0 = TMR1 register did not overflow

Note 1: PSPIF is reserved on PIC16F873/876 devices; always maintain this bit clear.

FIGURE 6.8   The bits in the PIR1 register (image courtesy Microchip Technology Inc.).

on, you can check pin 18 for the "heartbeat" to see if the PIC is executing its code.

The following code is the main loop of the program:

```
Main
    bsf HEART
    nop
    nop
    nop
    nop
    bcf HEART
    goto Main
```

REGISTER 2-4:   PIE1 REGISTER (ADDRESS 8Ch)

R/W-0	R/W-0	R/W-0	R/W-0	R/W-0	R/W-0	R/W-0	R/W-0
PSPIE[(1)]	ADIE	RCIE	TXIE	SSPIE	CCP1IE	TMR2IE	TMR1IE

bit 7   bit 0

bit 7   **PSPIE[(1)]**: Parallel Slave Port Read/Write Interrupt Enable bit
         1 = Enables the PSP read/write interrupt
         0 = Disables the PSP read/write interrupt

bit 6   **ADIE**: A/D Converter Interrupt Enable bit
         1 = Enables the A/D converter interrupt
         0 = Disables the A/D converter interrupt

bit 5   **RCIE**: USART Receive Interrupt Enable bit
         1 = Enables the USART receive interrupt
         0 = Disables the USART receive interrupt

bit 4   **TXIE**: USART Transmit Interrupt Enable bit
         1 = Enables the USART transmit interrupt
         0 = Disables the USART transmit interrupt

bit 3   **SSPIE**: Synchronous Serial Port Interrupt Enable bit
         1 = Enables the SSP interrupt
         0 = Disables the SSP interrupt

bit 2   **CCP1IE**: CCP1 Interrupt Enable bit
         1 = Enables the CCP1 interrupt
         0 = Disables the CCP1 interrupt

bit 1   **TMR2IE**: TMR2 to PR2 Match Interrupt Enable bit
         1 = Enables the TMR2 to PR2 match interrupt
         0 = Disables the TMR2 to PR2 match interrupt

bit 0   **TMR1IE**: TMR1 Overflow Interrupt Enable bit
         1 = Enables the TMR1 overflow interrupt
         0 = Disables the TMR1 overflow interrupt

Note 1: PSPIE is reserved on PIC16F873/876 devices; always maintain this bit clear.

FIGURE 6.9   The bits in the PIE1 register (image courtesy Microchip Technology Inc.).

In the code initialization, there was a #define statement that made "HEART" equivalent to "PORTC, 3." So this bit is set at the top of the loop. There are a several "nop" statements. These commands do nothing except consume two instruction cycles. An instruction cycle is four clock cycles. In this case, the clock frequency is 20 MHz and so the instruction cycle is 0.2 µs. Then PORTC bit 3 is cleared and the process is started over again.

Not terribly exciting, is it? The real action occurs in the *interrupt service routines* (IRS). There are four interrupts being used for the car: TMR0, TMR1, PSP, and RB0.

Interrupts

The following section discusses the components of interrupts.

Interrupt Decoding.   When an interrupt occurs, the program stops what it's doing, goes to memory address 0004h, and executes the command located there. As described above, this location has a "goto" statement pointing to the location of the ISR. When the program goes to the ISR, it only knows that an interrupt

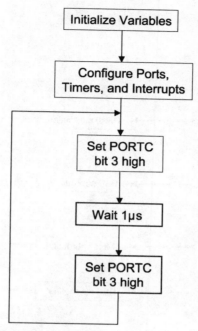

FIGURE 6.10 The main loop in the PIC program.

has occurred; it does not know *which* interrupt has occurred. So it must determine which interrupt sent it there. Fig. 6.11 shows the flow of the entire ISR.

The first thing that should happen during an ISR is context saving. When the interrupt is called, the program counter is saved on the stack so that when the ISR is over, the program can return to where it was when the interrupt occurred. However, the program counter is the only information that is saved. Context saving refers to temporarily storing key register contents in other memory locations in case those registers change during the ISR. The last thing done in the ISR is the restoration of register values. Typically, the two most important registers to save are the W and STATUS registers. This is done with the following code:

```
; Save STATUS and W registers
    movwf W_ TEMP          ; Copy W to TEMP register
    swapf STATUS,W         ; Swap STATUS to be saved into W
    clrf STATUS            ; Bank 0, regardless of current bank,
                           ; clears IRP,RP1,RP0
    movwf STATUS_ TEMP     ; Save status to bank 0 STATUS_ TEMP
                           ; register
```

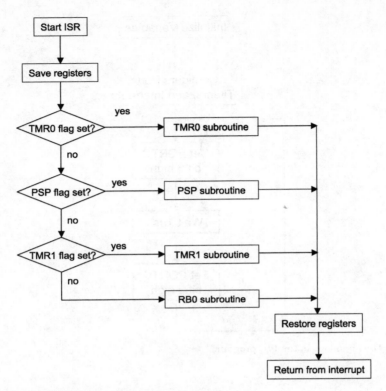

FIGURE 6.11   The flowchart for the interrupt service routine.

The last thing done before the ISR returns to the main loop is the restoration of these registers:

```
Ret
      swapf STATUS_ TEMP,0 ; Swap STATUS_ TEMP register into W
                           ; (sets bank to original state)
      movwf STATUS         ; Move W into STATUS register
      swapf W_ TEMP,1      ; Swap W_ TEMP
      swapf W_ TEMP,0      ; Swap W_ TEMP into W
      retfie
```

The "swapf" command is another way to move the contents of a file register to the working register. However, with this command, the upper and lower four bits, or nibbles, are exchanged when the contents are moved.

In this program, context saving isn't completely necessary because nothing really happens in the main loop. The STATUS and working registers are never accessed. However, in general, it is a good idea to save these registers. In the future, if more code is added to the main loop that does access these registers, there will be no loss of information when an interrupt occurs.

The next task is to figure out which interrupt occurred. This is done by checking each flag bit one at a time and "falling through" the code until you find the flag that is set.

```
; Determine which interrupt occurred

    btfss INTCON,T0IF ; See if the interrupt was TMR0
    goto PSPInt
    TMR0Int


---------------------------------------------
|                                           |
|              ...TMR0 subroutine...        |
|                                           |
---------------------------------------------

PSPInt
    btfss PIR1,PSPIF ; See if the interrupt was the PSP
    goto TMR1Int


---------------------------------------------
|                                           |
|               ...PSP subroutine...        |
|                                           |
---------------------------------------------

TMR1Int
    btfss PIR1,TMR1IF ; See if the interrupt was TMR1
    goto RB0Int


---------------------------------------------
|                                           |
|              ...TMR1 subroutine...        |
|                                           |
---------------------------------------------

RB0Int


---------------------------------------------
|                                           |
|               ...RB0 subroutine...        |
|                                           |
---------------------------------------------
```

The *bit test register file, skip if set* (btfss) command is used to check the individual interrupt flags. It works similar to an "if" statement in a higher-level programming language. First, it checks a particular bit, and if it is set, the next line of code is skipped. For example, at the beginning of the above code, the T0IF bit of the INTCON register is checked to see if it is set. If it is set, the TMR0 overflowed and caused the interrupt. So if this bit is set and the TMR0 interrupt did occur, the "goto" statement is bypassed and the TMR0 subroutine is executed. If this bit is clear, the TMR0 interrupt did not occur and the code bypasses the TMR0 subroutine, going on to check if the PSP interrupt occurred. This continues until the RB0 subroutine is reached. By the time the code reaches this point, the RB0 interrupt is the only choice left and so it is not necessary to check if that interrupt flag was set. Within each ISR subroutine, it is necessary to clear the flag bit so that it is not misread when the next interrupt is generated. Each subroutine is described separately in the following section.

TMR0.   TMR0 is used to generate the servo control pulse. Recall from Fig. 2.6 that a standard servo control pulse has a period of about 10 ms and the high time ranges from 1 to 2 ms. The flowchart in Fig. 6.12 shows the steps taken during the TMR0 interrupt service routine. The variable *steercnt* is a number between 0 and 156 that is received from the DSP that represents the desired steering angle. Values greater than 78 turn the wheels left, values less than 78 turn the wheels right, and 78 is center. When the program begins, the variables are initialized as $loop\_0 = 7$, $steercnt = 78$, $SERVOPIN = 0$, $hold = (1,0)$, and the prescaler is set to 32.

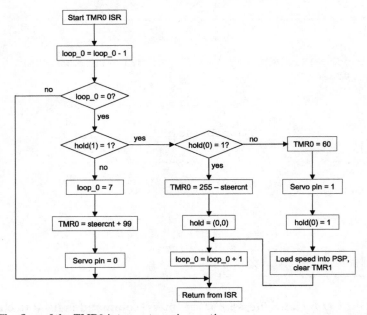

FIGURE 6.12   The flow of the TMR0 interrupt service routine.

Because of the specifications for a servo control pulse, there is always the same minimum amount of high time and low time. Taking advantage of this to improve resolution, the period of the servo control pulse is divided into nine sections, as shown in Fig. 6.13. The duration of sections 2 through 8 is fixed with sections 2 through 7 lasting 1.64 ms each and section 8 lasting 1.25 ms. The duration of sections 1 and 9 are variable, but their sum is fixed. The sum is 1 ms, and how long each section lasts depends on how far left or right the DSP wants the car to turn. This code creates a signal with a period of about 12 ms and a high time that varies from 1.25 to 2.25 ms. It was found, experimentally, that these high times gave the best range of steering for the car.

Each section terminates with a TMR0 interrupt during which a new value is loaded into the TMR0 register. The value that is loaded depends upon which section is terminating and the desired steering angle.

Also, during the TMR0 ISR, the TMR1 value is loaded into the PSP output buffer so that the value can be read over the data bus by the DSP. At that point, TMR1 is cleared. The output of the forward decoding D flip-flop is connected to the T1CKI pin on the PIC (see Chapter 4, "Environment Sensing"). TMR1 is configured as a counter and increments on the rising edge of the forward optical encoder pulses. So the value in the TMR1 register is the number of forward-encoder pulses counted in 12 ms. There are 512 slits in the optical wheel, so the rotational velocity can be calculated easily by the DSP. Every 12 ms, the car's speed is loaded into the PSP output buffer.

The following code is the TMR0 subroutine:

```
bcf       INTCON,2      ; Clear the interrupt


decfsz    loop_0,1
goto      Ret

btfss     hold,1        ; Hold LSB's - 10 - goto 1st high period
goto      soff          ; 11 - goto 2nd high (variable) period
                        ; 00 - goto low time
btfss     hold,0
```

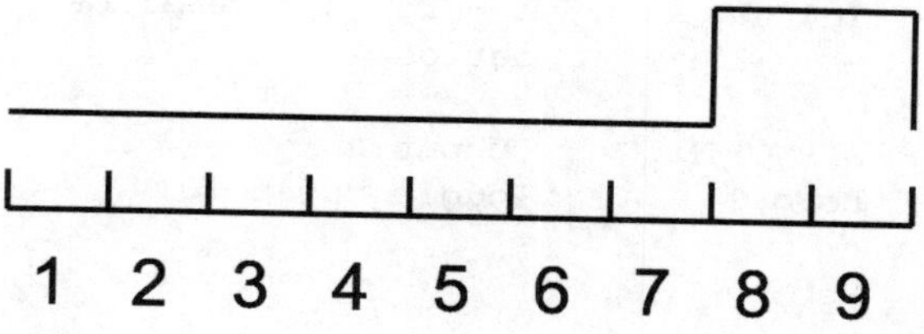

FIGURE 6.13 The servo control pulse is partitioned into nine sections to improve resolution.

```
            goto     shigh

            movfw    steercnt     ; Set TMR0 to put the steering
            sublw    d'255'       ; Pulse high for 255-steercnt ticks
            movwf    TMR0

            clrf     hold         ; Toggle the hold
            incf     loop_0,1
            goto     Ret
shigh
            movlw    d'60'        ; (255-60)*32*200ns = 1.25ms
            movwf    TMR0
            bsf      SERVOPIN     ; Set Servo output pin to high
            bsf      hold,0
            incf     loop_0,1

Feedback
            bcf      STATUS,C
            movf     r_speed,0    ; Compares Forward and Reverse Speed
            subwf    TMR1L,0
            btfss    STATUS,C     ; Is it in reverse?
            goto     Reverse
            movwf    PORTD        ; No - send speed to PSP
            goto     CLR
Reverse
            movf     TMR1L,0      ; Yes - set bit 7 high, send negative
            subwf    r_speed,0    ; speed to PSP
            movwf    PORTD
            bsf      PORTD,7
CLR
            clrf     r_speed
            clrf     TMR1L
            goto     Ret
soff
            movlw    d'7'
            movwf    loop_0
            movlw    d'99'

            addwf    steercnt,0   ; Set TMR0 to put the servo pulse
            movwf    TMR0         ; low for the remainder of the
                                  ; period

            bcf      SERVOPIN     ; Output a low
            bsf      hold,1       ; Toggle the hold

            goto     Ret
```

PSP. PORTD is configured as the PSP and is used to communicate over the data bus. When a PSP interrupt occurs, either a read or a write has been completed. During the ISR, the program checks to see if the input buffer is full. If it is full, then the DSP has just sent a steering or velocity command over the data bus.

Since the information must be sent from the DSP to the PIC eight bits at a time, the steering and speed commands are encoded. The DSP must provide a number between 0 and 156 for the steering angle, with 78 being the center. This determines the servo control pulse width and was described in detail in the previous section. For the speed command, the DSP must provide a 10-bit number to determine the PWM duty cycle and a bit for direction. Since the PSP is only eight bits, the information cannot be sent all at once.

As a result, the steering and speed information sent to the PIC must be broken into two 8-bit words that are sent separately. The command must tell the PIC whether it is receiving a steering or speed command, which 8-bit word it is, and what the speed or steering angle is. The encoding for the command is shown in Table 6.1.

In the PSP ISR, the program determines whether it just received a steering or speed command. It also determines if it received the high or low byte values. Once it has determined this, the values get loaded into the variable *steercnt* or the TMR2 register.

The following code implements the PSP subroutine:

```
PSPInt
    btfss   PIR1,PSPIF   ; See if the interrupt was the PSP
    goto    TMR1Int

    bcf     PIR1,PSPIF   ; Clear the interrupt
    btfss   TRISE,IBF    ; Check if data was written to PSP
    goto    Ret

    movf    PORTD,0      ; Read in the data from PSP
    movwf   cmdword
    movwf   temp1
    btfss   cmdword,7
    goto    LowByte
```

TABLE 6.1   Data Encoding for DSP/PIC Communication

Bit	Meaning
0-5	Steering or velocity value
6	0 = steering, 1 = speed
7	0 = low byte, 1 = high byte

```
        HighByte               ; Command Word is (1 x <6 MSBs of
                               ; steering/velocity)
            rlf     temp1,1
            rlf     temp1,1
            bcf     STATUS,C
            movlw   0xfc
            andwf   temp1,0
            btfss   cmdword,6

            goto    Hsteer

        Hspeed
            movwf   highspeed
            goto    Ret

        Hsteer
            movwf   steer
            goto    Ret

        LowByte
            btfss   cmdword,6
            goto    Lsteer

        Lspeed                 ; Command Word is <0 1 Vel1 Vel0 Dir x Vel3
                               ; Vel2
            movlw   0x03
            andwf   temp1,0
            addwf   highspeed,0
            movwf   CCPR1L
            movlw   0x30
            andwf   temp1,1
            movlw   0xcf
            andwf   CCP1CON,0
            addwf   temp1,0
            movwf   CCP1CON
            btfss   cmdword,3
            bcf     MOTORDIR
            btfsc   cmdword,3
            bsf     MOTORDIR
            goto    Ret

        Lsteer                 ; Command Word is <0 0 x x x x Str1 Str2
            movlw   0x03
            andwf   temp1,0
            addwf   steer,0
            movwf   steercnt
            goto    Ret
```

TMR1. TMR1 is used as a counter for the forward-encoder pulses and should never overflow. But if the car is going *really* fast, the register will overflow and the interrupt will occur. If this happens, the interrupt flag is simply cleared and the program returns to the main loop. For the timer to overflow, there would need to be more than 2^{16} encoder pulses in 12 ms. The encoder gives 512 pulses per revolution so an overflow would mean that the wheel would be spinning at more than 10,000 revolutions per second!

The following TMR1 subroutine simply clears the interrupt flag and returns to the main loop.

```
TMR1Int
    btfss   PIR1,TMR1IF
    goto    RB0Int

    bcf     PIR1,TMR1IF
    goto    Ret
```

RB0. The external interrupt, or RB0, is connected to the reverse-decoding D flip-flop. When this interrupt is generated, the variable *r_speed* is incremented. During the TMR0 ISR, this value is subtracted from the TMR1 register value that stores the forward-encoder count. Then the DSP knows if and how fast the car is going in reverse. During normal operation, the car is only going forward. However, this allows for the possibility of including reverse operation in the future.

The code for the RB0 subroutine simply increments the variable *r_speed*, which stores the reverse-encoder count:

```
RB0Int
    bcf     INTCON,INTF
    incf    r_speed,1
```

This completes the description of the PIC's program.

Programming the PIC

Now that we have the code that will make the PIC run the car, where do we go from here? There are two steps: First, the code must be assembled to create a file for download to the chip; second, the chip must be programmed with this file. The remainder of this chapter discusses the process of assembling the code and programming the chip.

The assembler has been mentioned several times in previous sections, but there has been no mention of what it does. Simply put, the assembler is a program that takes the code that you've written and translates it into something that the PIC can understand. The code you write is text, but the PIC's hardware can't understand text. The hardware only understands ones and zeros. The assembler's job is to convert the text file to a file containing the ones and zeros, called the executable machine code. This file is usually represented in hexadecimal format, which is why the result is often called a HEX file.

For example, to clear bit 0 of PORTC, you would put "bcf PORTC,0" in your code. The assembler translates this line into its opcode. The opcode for this command is 01 0000 0000 0111 in binary or 0×1007 in HEX. If you wanted to, you could write the entire program using the opcodes for each command. Then it would not be necessary to assemble the code since it would already be in the PIC's native language. However, writing and debugging this code would be extremely difficult!

MPLAB

Fortunately, Microchip provides IDE software called MPLAB. It is available for free from the company's Web site. MPLAB allows the user to write, debug, simulate, assemble, and download code all from the same software environment. The MPLAB IDE is shown in Fig. 6.14. MPLAB provides a text editor so that

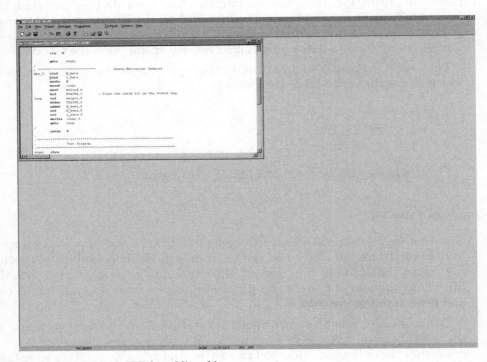

FIGURE 6.14   The MPLAB IDE from Microchip.

the user can create and edit source files such as assembly files that contain PIC code, or include files that contain configuration or register definitions. These files can be grouped together into a project. However, if you do not have several files comprising your source code, as is the case with the car code, it is easy to open the files individually. To open an assembly file, go to the File/Open menu and select the file. Make sure that the file has the ".asm" extension.

Once the file is opened, make sure that MPLAB knows which particular PIC the code is for. To do this, go to Configure/Select Device. This brings up the window shown in Fig. 6.15. Choose the particular PIC that you are using. In our case, it is the PIC16F874.

To assemble the code, go to Project/Quickbuild or Project/Build All. Which one is available depends on whether you have the files opened individually or as a project. After the build is complete, a separate output window appears that indicates whether the build was successful or if the code contains errors. If the build was unsuccessful, double-clicking on the error reported in the output window places an arrow next to the problem line in the source code. See Fig. 6.16. In this case, "movwf" was mistyped as "mowf" and the assembler mistook it for a label.

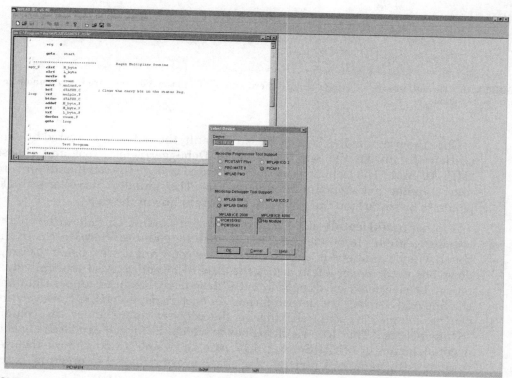

FIGURE 6.15   The dialog box to choose the correct PIC device.

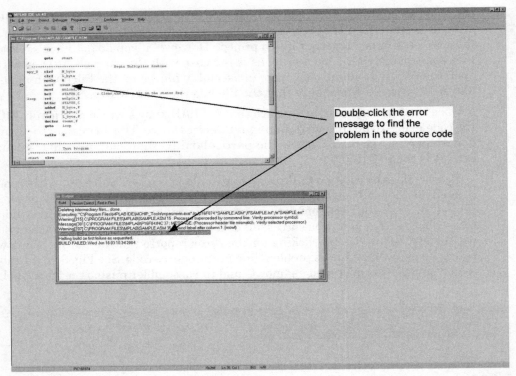

Double-click the error message to find the problem in the source code

FIGURE 6.16 The output window, after a build; if there were errors in the code, they are easily found using the output window.

Once the program has been successfully built, the code can be tested by simulation. To invoke the simulator, go to Debugger/Select Tool and choose "MPLAB SIM." The Debugger menu is activated to have the commands for running, stopping, and stepping through the code. The simulator also allows breakpoints to be set to stop the program at a certain line in the code.

Another useful feature of MPLAB is the watch window. This allows any of the register contents to be viewed. This is done by going to View/Watch. In the watch window that opens, choose the register you want to view using the drop-down box in the upper left and click the "Add SFR" button. Also, any individual bit can be viewed by selecting it from the drop-down box in the upper right and clicking Add Symbol. The default format for each register is HEX. However, this can be changed by right-clicking on the register name under the column "Symbol Name." This opens a dialog box in which the format can be changed to decimal, binary, or ASCII. See Fig. 6.17. Also, the watch window allows the user to change the value of a register. The changes are reflected in program as the simulation is running.

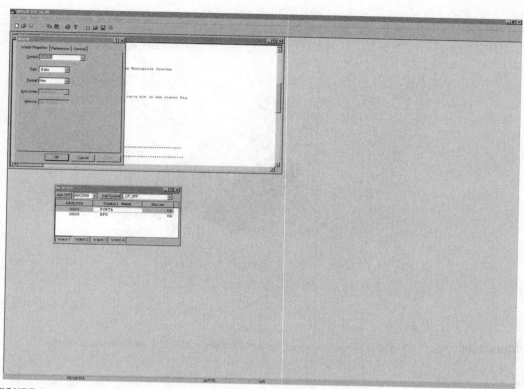

FIGURE 6.17   The watch window feature in MPLAB allows the values of registers and bits to be viewed while simulating the code.

Other features of MPLAB include a stopwatch that increments based on the PIC's oscillator frequency. This is useful for debugging time-critical code such as was written for the servo control pulse. There is also a built-in capability to change pin values. This is useful for simulating an external device connected to the PIC, such as the optical encoder whose value affects variables in the program. These features are easy to use and are well documented in the Help files that are included with MPLAB.

PIC programmers

There several ways to download the HEX file to the PIC. This section gives an overview of the different devices that are available.

There are two device programmers available from Microchip: the PRO MATE II and the PICSTART Plus. The PRO MATE II is a full-featured programmer that can program the entire family of PIC microcontrollers as well as many of

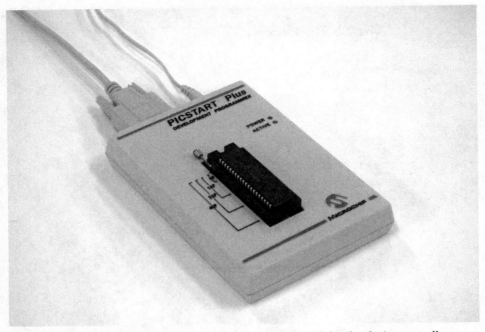

FIGURE 6.18 The PICSTART Plus programmer for use with the PIC family of microcontrollers.

Microchip's other products. The PICSTART Plus, shown in Fig. 6.18, does not support as many devices as the PRO MATE II, but it does support the entire family of PIC microcontrollers. Both programmers can be used from within MPLAB and connect to the PC via the serial port. Available from electronics suppliers such as Digikey, the PRO MATE II is the most expensive programmer, costing nearly $700, while the PICSTART Plus is about $200.

Another option is called the Warp-13 from Newfound Electronics. This programmer is similar to the PICSTART Plus and is compatible with most, but not all, of the PIC microcontrollers. The Warp-13 can be used from within MPLAB also. Available from retailers such as the ByteFactory (www.thebytefactory.com) and Junun Robotics (www.junun.org), this programmer costs about $100.

The aforementioned programmers are simple to use. They only require a power supply and a serial cable to work, and are fitted with sockets that allow for the easy insertion and removal of any size PIC. To download code from MPLAB using one of the above programmers, simply enable it by selecting PROMATE II or PICSTART Plus from the Programmer menu. Then compile the code by going to Project/Quickbuild or Project/Build All and press the Program button to download the code.

Many other programmers are not compatible with MPLAB. These programmers usually come with their own software that is designed specifically for use with that programmer. For the adventurous, countless Web sites provide schematics so that one can build a programmer from scratch. Many of these Web sites and the PIC datasheet itself provide details of how code is downloaded to the chip.

As was described in Chapter 2, the PIC is the slave processor that is the interface between the main processor and the motors. The next chapter describes the main processor in detail.

Microprocessor Control

The microprocessor is the brain of the computer. It collects all the data from the various sensors and, based on that information, decides how to move the car. This chapter describes the microprocessor in detail, from the program written for the car to the process of downloading code to the chip.

The microprocessor used on the car is the TMS320C6711 *digital signal processor* (DSP) from Texas Instruments. This DSP comes as part of a *DSP starter kit* (DSK) in which the DSP chip is located on a printed circuit board, so starting the DSP is immediate and easy.

Tutorial

We begin this chapter with a brief tutorial that gives the basics of how to write and download code to the DSP. This section takes you through the process of writing a simple program, compiling it, downloading it to the DSP, and finally downloading it to flash memory on the DSP's circuit board. To perform these steps, you will need the DSK with the TMS320C6711. The kit contains the DSP circuit board, a 5V power supply, a parallel cable, and the Code Composer Studio software. The software requires a computer with a 233 MHz processor, 600MB of free hard drive space, 64MB of RAM, and Windows 98 or later.

Using code composer

The first step is to install Code Composer on your computer. This tutorial assumes that the program was installed in the c:\ti directory. If you installed Code Composer in a different location, substitute the correct directory in place of c:\ti.

Once the computer is set up and able to communicate with the DSP, start Code Composer and go to Project/New. See Fig. 7.1. Give the project a name and choose the working directory. Make sure that the Project Type is executable and the Target is the TMS320C67XX. Then go to File/New/Source File to open a text editor window and type in the following:

```
#define EMIF_CE1 0x1800004
#define CE1_32    0xffffff23
#define IO_PORT   0x90080000
int i;

void main(void)
{
    while (1)
    {
        for (i = 0; i < 10000000; ++i)
        {
            // do nothing
        }
        *(unsigned volatile int *)IO_PORT = 0x0; /* Turn on all
user LEDs */
```

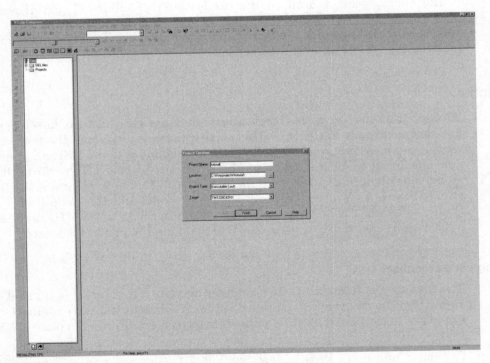

FIGURE 7.1   Creating a new project in Code Composer Studio.

```
        for (i = 0; i < 10000000; ++i)
        {
            // do nothing
        }
        *(unsigned volatile int *)IO_PORT = 0x07000000; /* Turn
off all user LEDs */
    }
}
```

Go to File/Save As and save this as a C source file. Next, create a new source file and type in the following:

```
MEMORY
{
    IRAM : origin = 0x0000, len = 0xFFFF
}

SECTIONS
{
    .boot_ load > IRAM
    .text       > IRAM

    .bss        > IRAM
    .cinit      > IRAM
    .const      > IRAM
    .far        > IRAM
    .stack      > IRAM
    .cio        > IRAM
    .sysmem     > IRAM
}
```

Go to File/Save As and save this as a TI command language file. Next, the following three files must be added to the project: the C source file and command file that were just created, and the C67XX library file. To add a file, go to Project/Add Files to Project and select the type of file that you want to add. First, choose C Source Files and select the C file created previously. Then, go back to Project/Add Files to Project, choose Linker Command File, and select the command file just created. Finally, go to Project/Add Files to Project and choose Object and Library Files. Then add the file *rts6701.lib*, which is located in c:\ti\c6000\cgtools\lib. Once this is done, save the project by going to Project/Save.

The next step is to compile the code by going to Project/Rebuild All. If the program was typed correctly and the proper files were added to the project, the window opened at the bottom of the screen should indicate that the build was completed with no errors or warnings. See Fig. 7.2. To load the program on the DSP,

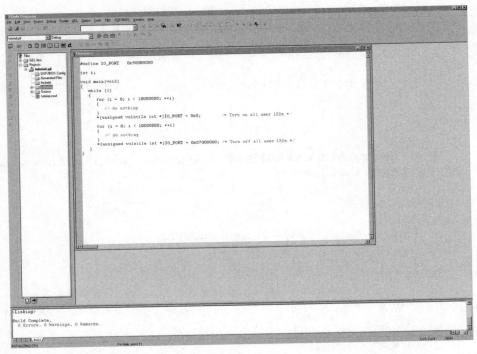

FIGURE 7.2   The build results for the tutorial project.

go to File/Load Program and select the ".out" file in the Debug subdirectory, which is located in the directory where the project was created. Once the download is complete, start the program by going to Debug/Run. When this program is running, the three *light-emitting diodes* (LEDs) on the DSP board will blink.

Using the Flash Programming Utility

If a program was downloaded with Code Composer, the program is lost as soon as the DSP is turned off or reset. However, if code is stored in the flash memory of the DSK board, it will begin executing as soon as the DSP is turned on or reset. A few more steps are necessary to load the code into flash memory. First, copy the file "boot.asm" from c:\ti\examples\dsk6711\board_util\post to your working directory, add it to the project in Code Composer, and rebuild the code. Then create the following text file:

```
Debug\tutorial.out
-a
-image
-zero
```

```
-memwidth 8
ROMS
{
    FLASH: org = 0x0000000, length = 0x10000, romwidth = 8, files
= {tutorial.hex}
}
```

Save this file as "hex.cmd" in the same directory as the previous project. Then copy the files from c:\ti\examples\dsk6711\board_util\flash\Host\Debug to this directory. Also copy the file "dsk6x11.cfg" from the c:\ti\cc\bin\BrdDat directory. From a DOS prompt, type the following two commands:

```
c:\ti\dosrun
hex6x hex.cmd
```

This creates the file "tutorial.hex" that will be loaded into the flash memory. Next, with the DSP board powered and plugged into the parallel port (make sure that the Code Composer program is closed), type the following at the DOS prompt:

```
flash tutorial.hex -fdsk6x11.cfg
```

This loads the tutorial program into the DSK's flash memory. Now, when the DSP is reset or turned on, *tutorial.hex* is loaded into the DSP's program memory and starts executing. Until the flash memory is overwritten with another file, the LEDs will blink upon powering the DSP.

Features of the DSP

The main processor on the FLASH car is a DSP, which is a processor that is specifically designed to perform multiply-and-add calculations very quickly and efficiently in hardware. This makes DSPs highly suitable for numerically intensive applications such as adaptive filtering, image processing, or robotics. The DSP was chosen for this application because of its computational power.

The DSP on the FLASH car is the TMS320C6711 from Texas Instruments. Like the PIC, the C6711 has its own instruction set. Unlike the PIC, which has 35 instructions, the C6711 has over 100 assembly instructions. This makes the DSP much more difficult to program using assembly. Fortunately, TI provides a C compiler for use with their DSPs. On the FLASH car, all of the DSP code was written in C using the Code Composer Studio *integrated development environment* (IDE). Code Composer is described in detail later in this chapter.

The C6711 is available as part of a DSK. The DSK also comes with a 5V power supply, parallel cable, the Code Composer Studio software, and a development board. The C6711 chip is on a development board, as shown in Fig. 7.3. This board allows for easy application development because it provides several connections that allow the DSP to access peripheral devices, including a parallel port for connection to a PC. In addition, the board contains 128KB of flash ROM and 4M×32-bit words of *synchronous DRAM* (SDRAM) that are accessible directly by the DSP.

The C6711 DSP is a 32-bit floating-point device operating at 150 MHz. The DSP's block diagram is shown in Fig. 7.4. There are 64K of internal memory available, and there is access to 64M×32 of external RAM. The chip also has a built-in boot loader so that programs can be loaded from *electrically erasable programmable ROM* (EEPROM) and run on the DSP. Additional peripherals include a serial channel, 16 *direct memory access* (DMA) channels, and 2 timers. The FLASH car utilizes the timers, external memory interface, and boot loader.

Timers. The C6711 has two 32-bit timers. Each can be configured to time events, count events, or generate signals. In addition, there are interrupts associated with each timer so that tasks can be performed at specified intervals. The timer block diagram in Fig. 7.5 shows its operation.

Each timer module consists of three 32-bit registers: the *control register* (TCL), the *period register* (PRD), and the *counter register* (CNT). The individual bits of each register can be changed or monitored depending on how the timer is operating.

FIGURE 7.3 The development board for the C6711 DSP.

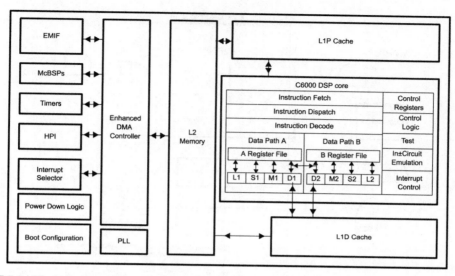

FIGURE 7.4 Block diagram of the C6711 DSP (from Texas Instruments).

The TCL determines which mode the timer operates in and monitors its status. The individual bits of this register control tasks such as halting or resetting the timer, setting the clock source as internal or external, or setting the timer to clock or pulse mode. Also, one bit, TSTAT, is used to hold the value that is output to the TOUT pin and can be used for generating signals.

The clock source can be set as either internal or external. When the internal clock source is used, the timer counter is incremented at 1/4 of the *central processing unit* (CPU) clock. In this case, the clock frequency is 150 MHz/4 = 37.5 MHz. If the internal clock is too fast, the user can provide an external clock to the TINP pin with an appropriate frequency.

For signal generation, the timer can operate in either clock or pulse mode. In clock mode, the TSTAT bit generates a signal with a 50 percent duty cycle and period determined by the value in the period register. In pulse mode, the TSTAT bit remains low until the timer overflows. Then TSTAT is set high for one or two clock cycles before being reset low.

The period and counter registers work in conjunction to determine the frequency of signals or interrupts generated. The PRD contains the number of input clock pulses to count. This number is written to the register and determines the frequency at which the timer overflows. The CNT increments every clock cycle until it reaches the value in the PRD, at which point it overflows back to 0. If the timer interrupt is enabled, an interrupt is generated when the CNT overflows.

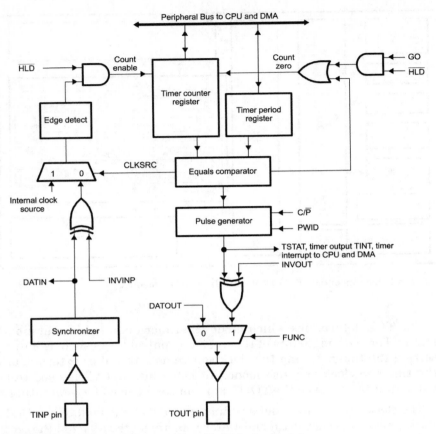

FIGURE 7.5   The DSP timer block diagram (from Texas Instruments).

External memory interface.   The FLASH car's various devices communicate with each other over the data bus as described in Chapter 4, "Environment Sensing." The DSP controls the data bus communication using its *external memory interface* (EMIF). The EMIF signals are shown in Fig. 7.6. The EMIF can be programmed to interface with synchronous burst *static random access memory* (SRAM), synchronous *dynamic random access memory* (DRAM), and asynchronous memory or devices. When configured to communicate with synchronous devices, the DSP handles much of the signaling so that the user does not have to program the device timing manually.

Data travels over the bidirectional 32-bit data bus (ED[31:0]) to the location specified on the 20-bit address bus (EA[21:2]) and the four chip select bits (CE[3:0]). There are also byte enable bits (BE[3:0]) that allow individual bytes or halfwords to be read from or written to the data bus. When the data bus is accessed, the appropriate signals are sent to the $\overline{AOE}/\overline{SDRAS}/\overline{SSOE}$,

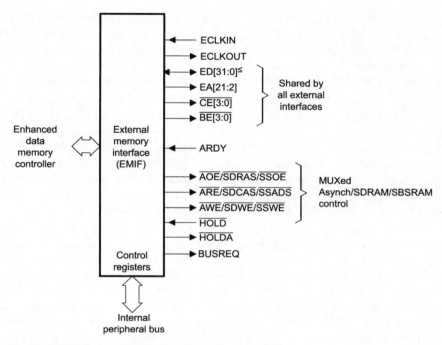

FIGURE 7.6   The EMIF control signals (from Texas Instruments).

$\overline{ARE}/\overline{SDCAS}/\overline{SSADS}$, and $\overline{AWE}/\overline{SDWE}/\overline{SSWE}$. The signaling depends on how the DSP has been configured to handle the interface. For example, if the DSP has been configured to communicate with an asynchronous interface, and a write operation is performed, the $\overline{AWE}$ is made low during the process while the $\overline{AOE}$ and $\overline{ARE}$ remain high.

Additional EMIF signals include *input* and *output clocks* (ECLKIN and ECLKOUT) for use with synchronous devices, an *asynchronous ready input* (ARDY) to insert wait states when communicating with slow memory or devices, *hold request* and *hold acknowledge* (HOLD and HOLDA) signals for external devices requesting data bus access, and a *bus request signal* (BUSREQ) that indicates bus access is pending or in progress.

Boot loader.   The boot loader is a program on the DSP that loads and executes code. When the DSP is powered on or reset, this program loads code from a particular location into the DSP's program memory and starts executing. The boot loader on the C6711 has four possible configurations: the host-port interface, 8-bit ROM, 16-bit ROM, and 32-bit ROM.

The *host-port interface* (HPI) is a parallel port on the DSP through which an external processor (such as a PC) can access the DSP's memory. The Code

Composer Studio software communicates with the DSP using the HPI. Code can be downloaded from a PC using Code Composer and can begin executing on the DSP. During the boot process, the DSP is held in reset while the host processor configures the DSP's memory. Once the host processor has completed initialization, the DSP is released from reset and begins executing the loaded code.

However, once the DSP is reset, the code loaded by the external processor is lost and must be reloaded. If the boot loader is configured to load the DSP's program from ROM, the DSP will begin executing the code stored in ROM whenever it is reset. Three different ROM widths are supported: 8-bit, 16-bit, and 32-bit. If the width is less than 32 bits, the DSP reads four 8-bit words or two 16-bit words at a time so as to utilize the entire 32-bit bus available for data transfer. When configured in this way, the DSP's program memory is loaded with the ROM data and the DSP begins executing upon device reset.

This section is intended to give a brief overview of the C6711 features that are used on the FLASH car. It is not meant to be an exhaustive description of the DSP's architecture or capabilities. Many other features available on the C6711 are useful in a wide variety of applications and can even be used for future expansion of the FLASH car. Complete details of the C6711 DSP hardware configuration are given in the *TMS320C6000 CPU and Instruction Set Reference Guide* and the *TMS320C6000 Peripherals Reference Guide*. Information on programming the DSP is in the *TMS320C6000 Assembly Language Tools User's Guide*, the *TMS320C6000 Optimizing C Compiler User's Guide*, and the *TMS320C6000 Programmer's Guide*. All of these documents are included with the DSP Starter Kit or are available from Texas Instruments.

The DSP Program Flow

The DSP is responsible for gathering information about the car, and determining how fast to drive and in what direction to steer. The program flow is shown in Fig. 7.7.

After initialization, the main loop of the program sequentially gets data from each peripheral sensor, processes that data, determines control inputs, and sends the control inputs to the PIC (whose programming was described in the previous chapter). The DSP program continues in this main loop until the car is turned off.

The next section describes all of the programming tasks in detail. The program is written in C and it is assumed that the reader has a working knowledge of the C language.

Program Details

The following section discusses program details such as initialization, calculation, automatic recharge, controllers, and conversion to PIC format.

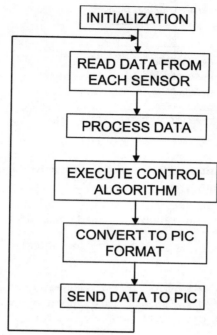

FIGURE 7.7 DSP program flow.

Initialization

There are four devices attached to the data bus with which the DSP communicates: *infrared* (IR) sensors, magnetic sensors, the PIC, and the *analog-to-digital* (A/D) converter. These are assigned the addresses as shown in Table 7.1.

The addresses are assigned to pointers as follows:

```
#define IR_ADDR  0xA0000000 // Address for the IR sensors
#define MAG_ADDR 0xA0000004 // Address for the magnetic sensors
#define PIC_ADDR 0xA0000008 // Address for the PIC
#define ADC_ADDR 0xA000000C // Address for the ADC

volatile int *ir = (volatile int *)IR_ ADDR;
volatile int *mag = (volatile int *)MAG_ ADDR;
volatile int *pic = (volatile int *)PIC_ ADDR;
volatile int *adc = (volatile int *)ADC_ ADDR;
```

To read from or write to one of the peripheral devices, it is only necessary to use the pointer variables (*ir, *mag, *pic, and *adc). The appropriate read, write, and chip select signals are generated automatically when the assignments are made.

TABLE 7.1   Memory Locations for the Peripheral Devices

Device	Address
Infrared sensors	0xA0000000
Magnetic sensors	0xA0000004
PIC	0xA0000008
A/D converter	0xA000000C

Also during the initialization, interrupts are disabled and the EMIF CE1 control bit is set as follows:

```
#define EMIF_CE1 0x1800004  /* Address of EMIF CE1 control   */
#define CE1_32   0xffffff23 /* reg to set CE1 as 32bit async */
*(unsigned volatile int *)EMIF_ CE1 = CE1_ 32; /* EMIF CE1
control, 32bit */

CSR=0x100;  /* Disable all interrupts            */
IER=1;      /* Disable all interrupts except NMI */
ICR=0xffff; /* Clear all pending interrupts      */
```

The first three lines configure the EMIF to communicate using a 32-bit wide asynchronous interface. The last three lines assign values to the *control status register* (CSR), *interrupt enable register* (IER), and *interrupt clear register* (ICR) so that interrupts are disabled.

Main program

The main loop of the program executes continuously until the car is turned off. This process is timed so that the precise sampling time of the car is known. At the top of the loop, timer 1 is started and at the end of the loop, after all of the data processing is complete, there is a wait loop that suspends the program until the timer has expired. At that point, the program returns to the top of the loop.

The registers associated with timer 1 are defined as follows:

```
#define TIMER1_CTRL 0x1980000  /* Address of timer1 control reg.
*/
#define TIMER1_PRD 0x1980004   /* Address of timer1 period reg.
*/
#define TIMER1_COUNT 0x1980008 /* Address of timer1 counter reg.
*/
```

The following code starts timer 1:

```
*(unsigned volatile int *)TIMER1_CTRL & = 0xff3f; /* Hold the
timer */
*(unsigned volatile int *)TIMER1_CTRL | = 0x200; /* Use CPU CLK/4
*/
*(unsigned volatile int *)TIMER1_PRD | = 0xffffffff; /* Set for
32 bit counter */
*(unsigned volatile int *)TIMER1_CTRL | = 0xC0; /* Start the
timer */
```

At the bottom of the loop, a wait loop is inserted that waits for the timer to expire before continuing:

```
while (timer1_read() < thresh)
{
    // do nothing
}
```

Here, *timer1read()* is a function that returns the value in the timer counter register and *thresh* holds the car's sample period in terms of the timer counter. As long as the timer counter is less than the threshold value, the sample period has not yet expired. This is done to ensure that the car's sampling rate is constant and does not vary depending on any conditional statements within the main loop.

Also, timer 0 is utilized to insert delays several times during the loop. The function *delay_msec()* is defined as follows:

```
void delay_msec(short msec)
{
    /* Assume 150MHz CPU, timer period = 4/150MHz */
    int timer_limit = (msec*9375)<<2;
    int time_start;

    timer0_start();
    time_start = timer0_read();
    while ((timer0_read()-time_start) < timer_limit);
}
```

The value passed to this function is the length of the desired delay in milliseconds and a timer value is calculated based on a 150 MHz clock. As with timer 1, *timer0_read()* returns the value in the timer 0 counter register. The desired delay is passed, in microseconds, to a similarly defined function *delay_usec()*.

Lateral displacement calculation.   The lateral displacement of the car from the center line is determined by both the IR and magnetic sensors located below the car (see Figs. 4.2 and 4.3). The IR and magnetic sensors give data in an identical format. The algorithm for each sensor type is the same, so only the IR sensors are discussed here.

To determine the lateral displacement, the data must be read in over the data bus, and then the data must be converted into actual distance, such as meters. The data is read in as follows:

```
input = *ir;
```

The execution of this line changes $A2, A3, !CE2, !ARE$, and $!AWE$ so that the address decoder enables the IR sensor buffer (as described in Chapter 4). Then the IR sensor data is put on the data bus and is read by the DSP into the variable *input*.

Each bumper has N sensors, so the DSP receives N bits of data for both the front and rear. These bits are normally 5V (logic 1), but in the presence of the line they become 0V (logic low). These bits are read in from both the front and the rear arrays at the same time, yielding an input of length $2N$ bits. The input is split into two variables, one for the front and one for the rear. The data bus is connected so that the leftmost bit is the *most significant bit* (MSB) and the rightmost bit is the *least significant bit* (LSB).

Once the DSP has the correct data from each array, it must convert this information into a distance. This distance represents how far the array's center has deviated from the line in the road. To determine the distance, the spacing of the sensors and the number of sensors must be known. These values are stored as constants in the C program.

The error distance is measured from the center of the array to the center of the active sensors. Fig. 7.8 shows two possible scenarios. In this figure, $N = 12$ and the data would be represented by 100111111111 for (a) and 110111111111 for (b).

To determine the actual distance in meters, the following algorithm is used. Starting with the LSB of the variable, the value of each bit is checked. If it is a 1, no line was detected by that sensor and the bit is ignored. However, if the bit is a 0, a line was detected by that sensor and so some calculation must be done. Each bit represents some distance from the center of the array, depending on the sensor spacing. Mathematically, this distance can be represented as

$$d = \left(\frac{N-1}{2} - i \right) \Delta x \tag{7.1}$$

where i is the bit's position (0 to $N - 1$) and Δx is the spacing between sensors. This distance is summed over all i and the result is divided by the total number of turned-on sensors as follows:

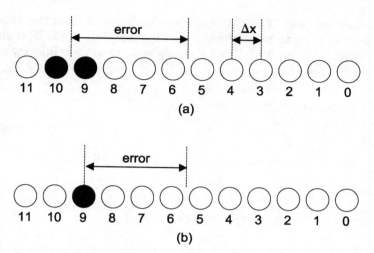

FIGURE 7.8   Two scenarios for the error measurement: (*a*) if two sensors detect the line and (*b*) if one sensor detects the line.

```
error = 0;
num = 0;
val = ((float)(bits)-1.0)/2.0;

for (i = 0; i < bits; ++i)
{
    mask = 0x0001<<i;
    current = mask & array;
    if (current == 0)
    {
        error = error+(val-i)*spacing;
        num = num+1;
    }
}

if (num == 0)
error = p_error;
}
else
{
    error = error/(float)(num);
}
```

Here, the variable *bits* is the same as *N* described above. The resulting error distance is positive if the bumper is to the left of the line. If none of the sensors are turned on, the car has lost the line and the error is assumed to be that which was calculated during the previous sample.

Speed calculation.   The PIC reads the car's speed from the optical encoder. During every pass through the program's main loop, the DSP requests the encoder count from the PIC. The following code reads the information from the PIC over the data bus:

```
encoder_count = *pic&0xFF;
sign_bit = encoder_count&0x80;
if (sign_ bit == 0x80)
{
    encoder_count = -(encoder_count&0x7F);
}
else
{
    encoder_count = encoder_count&0x7F;
}

actual_velocity = (float)(encoder_count)/30.54;
```

First, the value is read from the PIC and is ANDed with $0 \times FF$ so that only the eight LSBs contain the data. This is necessary because the PIC is an 8-bit processor and only sends 8 bits of information over the data bus. But the data bus is 32 bits wide and so the remaining 24 bits contain unknown values. The data is then stored in the variable *encoder_count*.

The value read from the PIC is the number of reverse encoder counts subtracted from the number of forward counts. The PIC and DSP both perform signed math, so if the car is moving backwards, the value read by the DSP will be negative. Thus, it is necessary to check the sign bit. The PIC uses the eighth bit as the sign bit while the DSP uses the thirty-second bit as the sign bit. The next nine lines convert the value in *encoder_count* from 8-bit to 32-bit representation by first masking out the eighth bit and storing it in the variable *sign_bit*. Then, if *sign_bit* is a one, then the value was negative and the magnitude of *encoder_count* is negated. Otherwise, the magnitude of *encoder_count* is stored back in itself, leaving a positive value.

Finally, the actual velocity of the car is determined by dividing *encoder_count* by 30.54. As described in the previous chapter, the PIC counts encoder pulses for 12 ms. Based on this counting window, the number of slits in the optical encoder wheel, and the diameter of the car's wheels, 30.54 is the number of encoder pulses that occur in 12 ms when the car is traveling at 1 m/s. So, *actual_velocity* is the car's speed in m/s.

Headway distance calculation.   The headway distance is determined using the IR range finder. To calculate how far an object is from the front of the car, the steps in Fig. 7.9 are done. The IR range finder outputs an analog voltage based on its distance from an object. This analog voltage is sent to the A/D con-

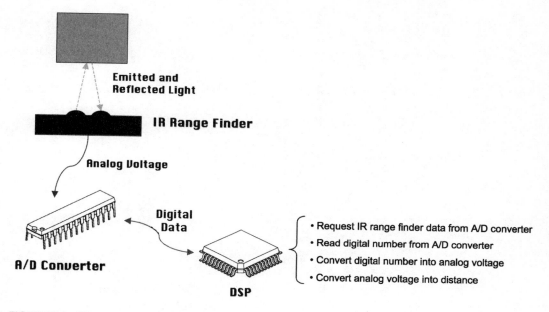

Emitted and
Reflected Light

IR Range Finder

Analog Voltage

Digital
Data

A/D Converter

• Request IR range finder data from A/D converter

• Read digital number from A/D converter

• Convert digital number into analog voltage

• Convert analog voltage into distance

DSP

FIGURE 7.9   The process of determining an object's distance from the car.

verter, which then outputs a 12-bit digital number over the data bus that is read by the DSP. The DSP program must perform the following four steps: request the IR range finder data from the A/D converter, read in the data, convert the data back to an analog voltage based on the A/D converter's characteristics, and convert the analog voltage to a distance based on the IR range finder's characteristics.

To request data from the A/D converter, the DSP must send configuration bits over the data bus. This is done by writing 0×42 to the data bus to indicate, among other things, on which channel to perform the conversion. After a delay of about 15 µs, the conversion has been made and the data is ready to be read by the DSP.

Next, the DSP must read the data. This is done by simply assigning the data at the A/D converter's memory address to a variable, as was done with the IR and magnetic sensors described.

Now that the DSP has the digital representation (a number between 0 and 4,095) of the IR range finder voltage stored in a variable, it must convert it back to an analog value. The configuration bits that were sent before the conversion were chosen so that 0×000 represents 0V and $0\times FFF$ represents 5V. The value then needs to be multiplied by $\frac{5}{4096}$, or 0.0012207, to get the analog voltage back. This is the voltage that was output by the IR range finder.

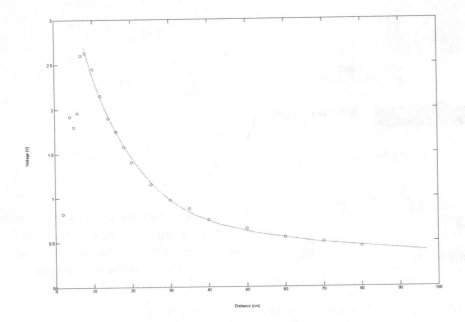

FIGURE 7.10 The circles show the sensor's data points as given in the datasheet, and the solid line is the curve fit to those data points.

Next, the voltage is converted back to distance. The response of the IR range finder is shown by the circles in Fig. 7.10. Using the mathematical software MATLAB, a curve was fitted to the data points and is described by

$$d = 715.77e^{-6.5284v} + 60.0672e^{-0.7573v} \qquad (7.2)$$

where v is the analog voltage output by the IR range finder and d is the distance in centimeters. The solid line in Fig. 7.10 shows the plot of the fitted curve.

As can be seen in the figure, the response of the IR range finder drastically changes when the object is closer than 8 cm. As a result, the fitted curve and the actual data points diverge when the object is very close. If the object is 5 cm away, the DSP thinks that it is 15 cm away. It is possible to account for the change in response by keeping track of which side of the peak the data points are on and using a different conversion accordingly. This was not done on the FLASH car because the distances used were all above 10 cm.

These steps are implemented by the following bit of code. The final step of the process is to divide by 100 to get the result in meters:

```
*adc = 0x42;
delay_usec(20);
input = *adc&0xFFF;
IR_ voltage = (float)(input)*0.0012207;
distance = 715.77*exp(-6.5284*IR_voltage)
+60.0672*exp(-0.7573*IR_voltage);
distance = distance/100.0;
```

Battery voltage and current calculation. As discussed in Chapter 4, the car monitors its own power level so that it can automatically recharge itself. The battery voltage and current are filtered through some circuitry and then sent to two different channels on the A/D converter. So, the method for reading the battery voltage and current is similar to that of the IR range finder voltage. They differ only in the configuration word sent to the A/D converter and the conversion formula for finding the original voltage. Because the voltage and current-sensing code is very similar, only the voltage sensing is discussed here.

Because there is some variation in the load placed on the battery as the car travels around the track, the voltage fluctuates. To get a better idea of the actual voltage level, the level is sampled every second and averaged over 30 samples. The following code implements the voltage sensing:

```
if (wait_count == 0)
{
    sum = sum-v_batt[sample_count];
    *adc = 0x40;
    delay_usec(20);
    input = *adc&0xFFF;
    battery_voltage = ((float)(input)*0.0012207/4.7+3.143)*2.0;

    v_batt[sample_count] = battery_voltage;
    sum = sum+v_batt[sample_count];
    v_batt_ave = sum/(float)(30);

    sample_count = (sample_count+1)%30;
}

wait_count = (wait_count+1)%2500;
```

The variable *wait_count* keeps track of when a sample is to be made. This variable is incremented each time through the main loop and a sample is only taken when *wait_count* is equal to zero. Since the main loop operates at 2500 Hz, *wait_count* becomes zero every second.

The value is read from the A/D converter as described in the previous section. First, a configuration is sent to the A/D converter to tell it to perform the conversion on channel 0. A delay occurs while the conversion is done. Then the DSP reads the value over the data bus. This value is converted back into the battery voltage by multiplying by the conversion factor and then by the inverse of the circuit's transfer function. This gives the instantaneous battery voltage.

The 30 voltage samples are stored in the *v_batt* array and *sample_count* is an integer between 0 and 29 that indicates the current sample location. The variable *sample_count* is incremented each time a sample is taken. The total of the 30 voltage samples is stored in *sum* and the averaged battery voltage that is used for calculation is this value divided by 30.

Automatic recharge.   The FLASH car is equipped with an automatic charging capability. This means that the car detects when it needs to be charged and pulls off the track into a charging bay. Automatic recharging takes advantage of the two different types of lateral displacement sensors, IR and magnetic. The car normally follows the white line around the track and then uses the magnets to pull off into the charging station. It is possible to have the car normally follow the magnets and then switch to the white line to pull off the track. The choice depends on the how the track is built.

The recharging algorithm has three modes: one when the cars first starts, one when the car is running normally, and one when the car needs to be recharged. When the car is operating normally, the recharging algorithm is not in effect. Fig. 7.11 shows the flow of the recharging algorithm. These steps are embedded in the main program along with the other tasks discussed in this section

When the car starts, it must be located in the charging bay for the automatic recharge to work properly. The startup mode is shown on the left side of Fig. 7.11. For the first few seconds, the car counts the number of distinct perpendicular lines it sees with its front IR sensor array. It is important that these lines are aligned so that the entire sensor array sees the reflection. During this time, the sensors used for lateral control are the magnetic sensors. After the startup time has elapsed, the car switches to a normal operating mode that uses the IR sensors for lateral control. Also, the number of perpendicular lines that were counted is stored as the car's bay number.

The car keeps track of the bay lines as it passes them while in normal operating mode. At any time, the car knows which bay pull-off is next, so if the car goes into recharge mode in between bay pull-offs, it does not start counting the perpendicular lines with the next one it sees (i.e., not bay number 1).

The car continues in normal mode until a battery recharge is needed. Then it switches to recharge mode to look for its own charging bay. When it finds the proper bay, the magnets direct the car off the main track into the charging station. There the car is turned off, and the battery is disconnected from the car's circuitry and connected to a charger using a relay. When charging is complete, the car's power is turned on and startup mode is active.

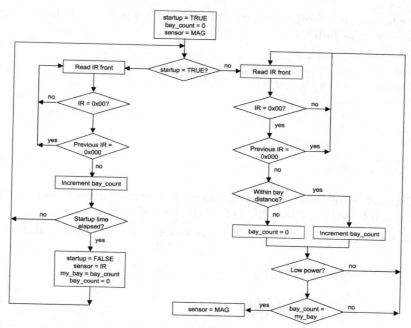

FIGURE 7.11   The flow of the automatic recharge algorithm.

Controllers.   Given the information from all of the various sensors, the controllers are the algorithms that bring the data together to determine the steering angle and speed inputs for the servo and motor, respectively. On the FLASH car there are several different controllers with different objectives: the lateral controller, the automatic cruise controller, the steering controller, and the speed controller. The theory and design of these controllers are discussed in detail in Part II, "Theory of Mobile Robots," of this book.

Conversion to PIC format.   The final task in the main loop is to send the steering and speed commands to the PIC. The output of the controllers described previously is a steering angle and a motor velocity. Specifically, the steering controller outputs a value between 0 and 156, where 0 and 156 are full left and right, respectively, and 78 is center. The PIC requires this range. The speed controller outputs a value between −1,023 and 1,023. This value must be converted to a direction and magnitude for the PIC to generate the required *pulse width modulation* (PWM) signal.

When sending information to the PIC, the DSP includes not only the actual value, but also what is being sent. All of this data requires more than the 8 bits available on the PIC's parallel slave port, so it is necessary to divide the information into 8-bit words, as indicated in Table 6.1. For the steering angle, this is done as follows:

```
output = (angle>>2)& 0x3F;
output = 0x80|output;
*pic = output;

delay_usec(10);
output = angle&0x03;
output = 0x00|output;
*pic = output;
```

The delay function between the two data writes ensures that the first write has been handled by the PIC before the second is sent.

A similar method is used to send the motor velocity information to the PIC:

```
if (PIC_velocity > 0)
{
   dir = 0;
   magnitude = PIC_velocity;
}
else
{
   dir = 8;
   magnitude = 1023+PIC_velocity;
}

/* send velocity high byte */
delay_usec(10);
output = 0xC0|(magnitude>>4)&0x3F;
*pic = output;

/* send velocity low byte */
delay_usec(10);
v10 = magnitude&0x03;
v32 = (magnitude>>2)&0x03;
output = 0x40|v10|dir|0|v32;
*pic = output;
```

The value of *PIC_velocity* is between −1,023 and 1,023, where −1,023 is reverse full speed and 1,023 is forward full speed. This is converted into a magnitude and direction, encoded as in Table 6.1, and sent to the PIC as two words.

Programming the DSP

A DSP program generally consists of C source files, assembly source files, and library files. The C source files contain the high-level, user-written code that runs on the DSP. The C source files are translated into assembly code (which is low-level code specific to the DSP) by the C compiler. There may be instances where it is appropriate to write some bits of code in assembly rather than C (for instance, if optimized code is desired). The assembler takes assembly files (those generated by the C compiler as well as user-written assembly code) and creates object files containing machine code, the DSP's native language. The object files and library files (containing run-time support for the DSP) are linked together using a command file that tells the linker where to place specific sections of the code. The output of the linker is an executable file known as a *common output file format* (COFF) file. This COFF file can be either downloaded to the DSP's memory through the HPI or converted into HEX format and downloaded to the onboard flash memory for boot loading. See Fig. 7.12. All of the tools necessary to complete this process with the C6711 DSP are included with Code Composer Studio from Texas Instruments.

Code Composer Studio

Code Composer Studio is to the DSP what MPLAB is to the PIC. It is a Windows-based program that allows the user to write, debug, simulate, build, and download code from the same environment. However, unlike MPLAB, Code Composer is not available free but comes with the purchase of the DSP Starter Kit.

Code Composer allows the user to create projects containing all of the files associated with the code. In the beginning of this chapter, the tutorial demonstrated the creation of a project. A project is saved with a .pjt extension and contains all of the files, as well as software tool options and dependencies.

The next step of the tutorial is the build. During the build, the code is compiled, assembled, and linked to create the COFF files. The compiler translates the source code into assembly and creates blocks of code known as sections. There are two kinds of sections: initialized and uninitialized. Initialized sections contain data while uninitialized sections are empty with space reserved for variables. Table 7.2 summarizes the sections created by the compiler and their contents. The .cinit, .const, .switch, and .text sections are initialized while the .bss, .stack, and .sysmem sections are uninitialized. Next, the assembler translates the assembly code into object files.

The final part of the build process is the linker. The linker combines all object files into an executable file that can be loaded into the DSP's memory and run.

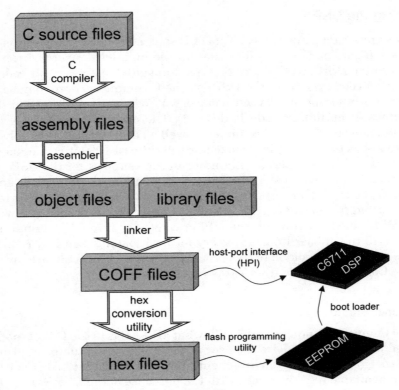

FIGURE 7.12   The code development process for the DSP.

TABLE 7.2   Sections Created by the C Compiler

Name	Contents
.cinit	Tables for explicitly initialized variables
.const	Explicitly initialized constant variables
.switch	Tables for large switch statements
.text	Executable code
.bss	Global and static variables
.stack	Stack memory
.sysmem	Dynamic memory

During this process, all of the sections created during the compile are located into specific memory locations according to the linker command file. The linker command file contains the linker options and memory location definitions.

Once the build is completed successfully, the program can be downloaded to the DSP's memory. With the program on the DSP and ready to execute, it can be debugged by stepping through the code and utilizing the breakpoints and watch windows that are available in Code Composer.

The watch windows allow the user to view and change variables while the program is executing. To invoke the watch window, go to View/Quick Watch to bring up the window shown in Fig. 7.13. Type the name of the variable you wish to view and click Add to Watch. The window in Fig. 7.14 will appear. The display format can be changed in the far right column to hexadecimal, decimal, or binary. To add another variable to the list, simply click in an empty row in the first column and type the desired name.

The watch window becomes very useful when the program has breakpoints. A breakpoint is a point in the program at which execution stops. To set a breakpoint, open the source code by double-clicking the file name. Then, double-click in the left margin of the desired line of code. A red dot appears to indicate that a breakpoint is set. When the program is run, it executes up to the line with the first breakpoint, but that line itself does not execute. The next line to execute is noted by a yellow arrow in the left margin. If any variables have changed value during the execution to that point, the value in the watch window is red. See Fig. 7.15.

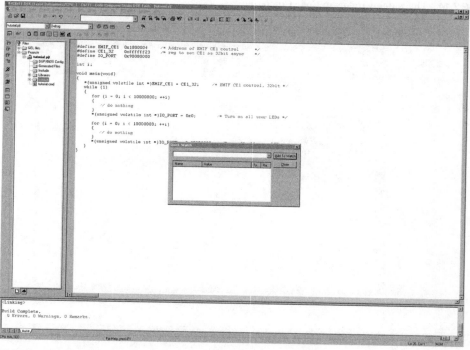

FIGURE 7.13   The window adding a watched variable.

FIGURE 7.14   The watch window in Code Composer.

FIGURE 7.15   The use of breakpoints in Code Composer.

In addition to using breakpoints, several methods are available in the Debug menu for executing a program. These include step into, over, and out; animate; run free; and run to cursor. Stepping through a program executes one line at a time. When a function call is encountered, there are two options: step into or step over. The step into command takes the program into the function so that each line in the function can be executed. The step over command executes the function at once and also pauses the execution again when the program returns. The step out command is used within a function. It causes the remainder of the function to execute and pause execution upon return to the main program.

The animate command is the same as the run command except that the program window is "animated;" the line of code being executed is highlighted. Run free executes the code as the run command does, but the breakpoints are disabled. This is useful because one can execute the code without stopping to add and remove the breakpoints. The run to cursor command executes the code up to the line containing the cursor. This allows sections of code to be executed without having to set breakpoints on each desired stopping line.

This section was intended to provide an overview of Code Composer's capabilities. Code Composer is a full-featured IDE and has many more assets not covered here. Complete documentation on all of the features is given in the on-line help files.

References.   The following is a list of useful documentation on the C6711 DSP and Code Composer Studio. These documents (and many more) are provided with the DSP Starter Kit in PDF and are also available from the Texas Instruments Web site.

Code Composer Studio Getting Started Guide, Literature Number: SPRU509C.

TMS320C6000 Programmer's Guide, Literature Number: SPRU198.

TMS320C6000 Technical Brief, Literature Number: SPRU197.

TMS320C6000 CPU and Instruction Set Reference Guide, Literature Number: SPRU189.

TMS320C6000 Peripherals Reference Guide, Literature Number: SPRU190

This part of the book has described the FLASH car's hardware and software in detail. With this information, the reader is well on the way to building a mobile robot capable of sensing its environment and driving itself. Starting with the next chapter, the theory of mobile robots is introduced. The remainder of the book deals with the modeling and control of the robots. This information is intended for those who are interested in the mathematics behind mobile robots or who are conducting research in this area.

Part
2

Theory of Mobile Robots

Modeling and Control Basics

Starting with this chapter, we take a look at the theoretical aspects of mobile robots. In particular, this part deals with the mathematics used for modeling and control.

This chapter makes the transition from Part I, "Hardware Implementation," to the theoretical discussions in the remaining chapters. It starts with a brief, general introduction to the concepts of modeling and control and assumes no prior knowledge in this area. The last part of this chapter focuses on how these ideas are related to mobile robots. It serves as an introduction to the ideas and tools in this rich area of research.

The aim of this chapter is to answer the following questions:

- What is modeling?
- Why is it useful?
- What is control?
- How are modeling and control related?

What Is Modeling?

The Merriam Webster dictionary defines a model as "a miniature representation" or "an example for imitation or emulation." We are all familiar with model airplanes or model students. The way we use the word "model" in everyday language gives us some idea of how the word is used in science. One scientific definition of a model, as given in *An Introduction to Mathematical Modeling* (Bender 1978), is a characterization that "mimics the relevant features of the situation being studied." One model that many people are familiar with is the law of supply and demand. Simply stated, as the demand for a product increases, the supply decreases, and thus its price increases. One can use this model to predict the price of a product based upon consumer demand.

Predicting what will happen to the system based on known variables is a major outcome of modeling. Whether you are predicting the weather for tomorrow based on measured variables such as temperature and barometric pressure, predicting a stock price based on its past performance, or predicting the location of a robot based on the speed and angle of the wheels, the idea is the same. A model tells you how something will behave.

However, what the model tells you may not be true to the real system. If it were, meteorologists would never be wrong and no one would lose money in the stock market. Models can never *exactly* capture every aspect of a system. There is always some modeling uncertainty, and if one tried to account for every aspect of the actual system, the model would be hopelessly complicated. This is the trade-off between accuracy and simplicity that one encounters when given a modeling task. A simple model is easy to work with because there are fewer variables to use (and thus measure), but with such a model the predictions are not very accurate. A complicated model may predict behavior very well, but it may require three weeks of computing time to arrive at that very accurate solution.

When formulating a model, the modeler must be aware of what purpose it will serve. For example, if an airplane model is going to be used in a flight simulator, it should mimic the actual airplane as closely as possible so that the simulator feels like the real flying experience. If the model is going to be used for designing a feedback controller, a simpler model may be sufficient if the controller can be designed to be robust to modeling uncertainty. The different objectives impose different requirements on the model itself.

The next question is then, what are the reasons for modeling a system? As described in *Case Studies in Mathematical Modelling* (James 1981), there are three basic uses for a model:

- **Understanding systems.** Scientists have developed models that try to explain the origin of the universe so that we can understand how it began.

- **Predicting the system's behavior in the future, or in different circumstances.** Meteorologists use models to obtain a forecast for the tomorrow's weather.

- **Controlling the system.** Control engineers use models of robots to develop algorithms for automating the robot's movement.

Modeling is an area of research in itself that spans many different disciplines. From robotics to biology, from manufacturing to economics, and everywhere in between, researchers are actively trying to develop better ways to describe a system's behavior. To get an idea of the breadth of this area, see *Case*

Studies (James 1981) or *Applied Mathematical Modeling* (Shier 2000). These books give many examples of areas where modeling is used, and they provide much detail.

What Is Control?

The scientific definition for control is closer to the definition we use in everyday life than it is with modeling. Simply stated, control means making a system do what we want it to. This idea is classified into two types of control: openloop and closed-loop.

Open-loop control

The idea of open-loop control is illustrated in Fig. 8.1. In this scenario, the controller is given a reference signal. Based on this signal and a model of the system to be controlled, it determines what control input must be sent to the system (typically called the *plant*). The controller has no idea what the plant is actually doing; it only has predictions based on the model. In fact, the control calculation can be done *offline*, when the system is not even operating.

As a simple example, consider a DC motor that must maintain a specified angular velocity, w. The plant may consist of the motor and an H-bridge, in which case the plant input would be the current supplied to the H-bridge. The openloop controller would have w as its reference input. Based on a motor model that relates current and angular velocity (such a model can be derived from the physics of the motor, or an experiment), one could determine how much current is necessary to achieve the desired angular velocity. So, supplying the necessary current would cause the motor to turn with w . . .

. . . or would it? Suppose one of the parameters used in the model was inaccurate, or a different motor is used. In either case, the relation between the current and angular velocity is different from the model used to determine the necessary current. It may be that the criterion was an angular velocity of *approximately w* and the differences are so small that they don't matter. If the requirements are stricter, this problem could be overcome by testing each motor separately and calibrating the input to achieve the desired angular velocity. However, if these systems were being mass-produced, it would be advantageous if each unit didn't need to be adjusted individually. Another problem is that as

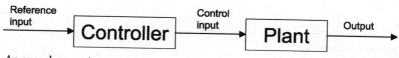

FIGURE 8.1 An open-loop system.

the motors wear over time, the relation between current and velocity can change. So, even if the current used today causes the motor to turn at the desired speed, will that same current work a year from now?

Although these issues may not be problematic for a particular application, there is something more serious to consider. Suppose someone puts his or her hand around the motor shaft and, due to the increased load, the motor slows down. With open-loop control, there is no way of knowing that this is happening, and while the same current is supplied, the shaft speed drops below an acceptable level. This is an example of a disturbance that can adversely affect the performance of an open-loop system.

There are so many potential problems that can cause the open-loop control system to perform poorly: unmodeled dynamics, parameter uncertainty and drift, and disturbances. What to do? One answer is to close the loop.

Closed-loop control

The block diagram for a closed-loop control system is shown in Fig. 8.2. As with open-loop control, the controller is given a reference signal that indicates the desired output of the system. But in closed-loop control, the controller has more information; it knows what the plant is actually doing. It can then compare the actual performance with the desired performance and adjust the control input accordingly. For closed-loop control, it is necessary to have sensors measuring performance of the system so that the information can be fed back to the controller.

Going back to the motor example, by placing an optical encoder (see Chapter 2, "Overall System Structure,") on the motor shaft and sending its output to the controller, the desired and actual speeds could then be compared. If this were done with various motors whose parameters were slightly different, the controller would compensate, causing the speed to be the same for each motor. Also, if a load were placed on the motor shaft, the controller would know that the angular velocity had slowed, so it would send more current through the H-bridge.

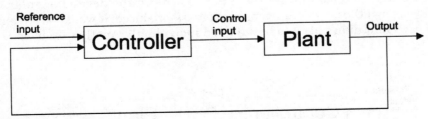

FIGURE 8.2 A closed-loop system.

The addition of the optical encoder and a sophisticated controller would make the system more complicated, but would also provide more robust performance.

Closed-loop control, also known as feedback control, improves a system's performance at the expense of added complexity. Not only are sensors necessary, but more time must be spent designing and building the controller. However, in many cases, the cost of additional hardware and time are more than worth the effort, due to the superior performance of the resulting system.

Feedback controllers can be implemented in a variety of ways. The driver in a car is a feedback controller. By taking in information about the roadway and surrounding cars, the driver makes adjustments to the speed and steering to avoid accidents and to arrive safely at the destination. In many instances, the controller is implemented by a computer or microprocessor and is written in a programming language such as C or C++. The sensor information is sent to the computer and is interpreted, filtered, or otherwise manipulated into a form that is useful for the controller. The controller itself is simply a formula that calculates a plant input based on the information it has received.

A Little Bit of Math

Up until this point, the discussion has been qualitative. However, modeling and control theory are highly mathematical fields. Mathematics provides a language that can precisely and concisely formulate the problems and their solutions. This section provides a brief introduction to the notation that is used in these areas.

Modeling involves understanding how variables are related to each other. For example, in a simple DC motor model there are two variables, torque (τ) and current (I). They are related by a constant

$$\tau = KI \tag{8.1}$$

This relationship is algebraic because the variables are not changing with time. Such a relationship is *static*. In many cases, we are interested in how the variables are changing with time, or how the system evolves. For example, how does a mobile robot's position change as it is given steering and velocity inputs? These are dynamic systems and their models are in the form of differential equations (for continuous time systems) or difference equations (for discrete time systems).

It is common to denote a system's variables (also called state variables) by x, its inputs by u, and its outputs by y. A system can have any number of state variables, so in general x is a vector, $(x_1, x_2,..., x_n)^T$, where n is the number of states in the system. Similarly, u and y are vectors in general. How the states change with time is given by $\frac{dx}{dt}$ and is usually denoted by $\dot{x}$. An example of a simple system with two state variables and one input would be

$$\dot{x}_1 = x_2$$

$$\dot{x}_2 = u$$

$$y = x_1 \tag{8.2}$$

This is a linear system because it can be represented by the form

$$\dot{x} = Ax + Bu$$

$$y = Cx + Du \tag{8.3}$$

where A, B, C, and D are appropriately sized matrices. For the example given by (8.2), the matrices are

$$A = \begin{bmatrix} 0 & 1 \\ 0 & 0 \end{bmatrix} \tag{8.4}$$

$$B = \begin{bmatrix} 0 \\ 1 \end{bmatrix} \tag{8.5}$$

$$C = \begin{bmatrix} 1 & 0 \end{bmatrix} \tag{8.6}$$

$$D = 0 \tag{8.7}$$

An example of a nonlinear system is

$$\dot{x}_1 = x_1^2$$

$$\dot{x}_2 = x_1 u$$

$$y = x_1 \tag{8.8}$$

This system is nonlinear because it cannot be represented in the form (8.3). Linear system theory is quite mature and many controllers for linear systems have been developed. However, most real-world systems (such as mobile robots) are nonlinear. In some cases, nonlinear systems can be linearly approximated, so linear controllers work quite well. But many times the system must operate outside its linear range, or the system may be too nonlinear for these methods to work. As a result, the study of nonlinear control theory is necessary to achieve good performance in such systems. For those interested in this area, an excellent and readable introduction is *Applied Nonlinear Control* (Slotine 1991).

Mobile Robot Modeling and Control

For the rest of this chapter, the discussion focuses on car-like mobile robots. The following sections provide a general view of modeling and control in this particular area. The details are provided in the following chapters.

Kinematic vs. dynamic modeling

There are two ways to model the mobile robot: kinematically or dynamically. In a kinematic model, only the movement of the vehicle is considered. The model is derived using the so-called nonholonomic constraints inherent in the vehicle. These constraints are described in detail in the next chapter. The kinematic model is fairly easy to derive and is simple in form. However, this model fails to account for acceleration and can differ greatly from the actual movement when the vehicle's handling limits are approached. Because of its simplicity, many researchers use the kinematic model and place the emphasis on designing robust controllers.

On the other hand, dynamic modeling accounts for the properties of the vehicle related to its acceleration such as mass and center of gravity. This type of modeling is much more true to the actual behavior of the vehicle, but the resulting system representation is much more complicated. See "Design and Analysis of Combined Longitudinal Traction and Lateral Vehicle Control for Automated Highway Systems Showing the Superiority of Traction Control in Providing Stability During Lateral Maneuvers" (Kachroo and Tomizuka 1995) and "Microprocessor-Controlled Small-Scale Vehicles for Experiments in Automated Highway Systems"(Kachroo 1997) for examples.

Control objectives

What would you like the robot to do? The answer to this question is the control objective. In terms of the mobile robot's motion in an obstacle-free environment, the control objective is classified into three categories: point-to-point stabilization, path following, and trajectory tracking. See Fig. 8.3.

The goal of point-to-point stabilization is for the vehicle to move from point A to point B without restrictions on its movement. The objective is defined as the car's final location and orientation. However, the objective may be further refined by requiring that the car move to the final position using a minimum amount of energy or traveling a minimum distance.

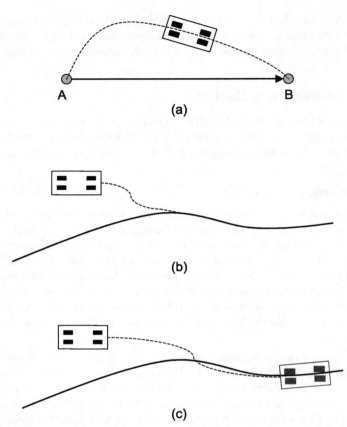

FIGURE 8.3 Three control objectives for a mobile robot: (*a*) point-to-point stabilization, (*b*) path following, (*c*) trajectory tracking.

With path following, the car must move along a geometric path with no restrictions on the speed. If the car senses that it is off the path, it must steer itself back onto it. As with point-to-point stabilization, additional requirements may be that the car move back onto the path in minimum time or distance.

Trajectory tracking is the most restrictive of the three control objectives. It is similar to path following, except the car must follow a path at a given (not necessarily constant) speed. Imagine that there is a virtual vehicle driving along the path that must be followed. The controller must drive the car so that it "catches up" to the virtual vehicle and stays with it.

In the following chapters, the goal for the car's movement is path following. The vehicle must sense its position with respect to the path and return to the path if it is off course.

Control approach

The control of a mobile robot can be viewed as a hierarchical system of three controllers: the motion planner, the path follower (or point-to-point stabilizer, or trajectory tracker), and the actuator driver. See Fig. 8.4.

At the highest level is the motion planner. At this level, it is determined what path and what velocity profile the robot is to follow. At least part of this motion planning must be done offline to determine the ultimate goal of the robot's motion. In this case, the path to follow and velocity are determined beforehand (e.g., a line to follow is painted on the road and the robot must maintain a safe speed at all times). However, online motion planners are available so that the robot can modify its movement based upon changing environmental factors. See "A Safe and Robust Path Following Planner for Wheeled Robots" (Lambert 1998) for examples.

Once the mobile robot knows where it must go, the controller at the next level takes on the task of making it get there. At this level, the actual position and velocity of the robot are measured and compared to the desired position and velocity (as determined by the motion planner). Based on the errors between the

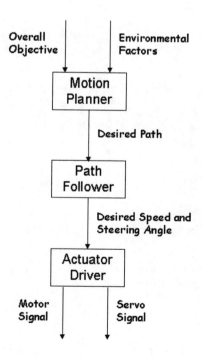

FIGURE 8.4   The control hierarchy of a mobile robot.

actual and desired state, the controller determines what steering or velocity inputs are necessary to achieve the desired position and velocity.

At the lowest level are the actuator drivers. These controllers receive as input steering and velocity commands from the previous level controller and determine what inputs to the motors are necessary to achieve the desired position or rotational speed. In other words, the inputs are the desired steering angle and velocity from the path follower. The actual steering angle and wheel speed are measured and compared with the desired values. This controller determines the necessary motor and servo signals to achieve the desired values.

Because of this control structure, the different levels of controllers can be decoupled and designed separately. The mobile robot models used for path following are kinematic. Only the movement of the robot is modeled, and the dynamic effects such as mass or center of gravity are not. The models used for the actuator drivers are dynamic. The hierarchical structure allows the robot dynamics to be compensated for at the lowest level.

Now that you understand the idea behind robot modeling, the next chapter gives the details of the mathematics.

Mathematical Modeling

The previous chapter gave a broad description of what the mathematics behind modeling and control theory is all about. Now we turn to the details. In this chapter, several models for the mobile robot are derived. In the next chapter, these models will be used to design controllers.

Before developing the models used for the mobile robot, it is appropriate to define the framework to be used throughout the rest of this chapter. In the following sections, three different frames of reference are used in describing the model: the global frame, F_g; the mobile robot frame, F_m; and the camera frame, F_c. A top view of the mobile robot is given in Fig. 9.1. The global frame, F_g, is fixed and the mobile robot frame, F_m, is attached to the robot and moves about in the global frame's (z,x) plane. The orientation of F_m is such that the linear velocity of the robot is along the z_m-axis. It is assumed that robot's environment is such that there is no movement in the y-direction, and that the y-axis and y_m-axis remain parallel at all times.

The camera frame, F_c, is attached to the mobile robot as shown in Fig. 9.2. This frame is chosen so that the origins of F_m and F_c are the same and that the x_m-axis lies along the x_c-axis. The camera is mounted at height h above the (z,x) plane. If, on the actual robot, the origins do not coincide, then a point in the actual camera frame can be transformed into F_c by a simple translation. Also, it is assumed that the camera is tilted downward so that α, the angle between the z_c-axis and z_m-axis, is positive. For implementation, it is required that $\alpha \in (0, \frac{\pi}{2})$ if the camera is to view the area directly in front of the robot.

Kinematic Modeling

The first and simpler model we deal with is the kinematic model. This type of model allows for the decoupling of vehicle dynamics from its movement.

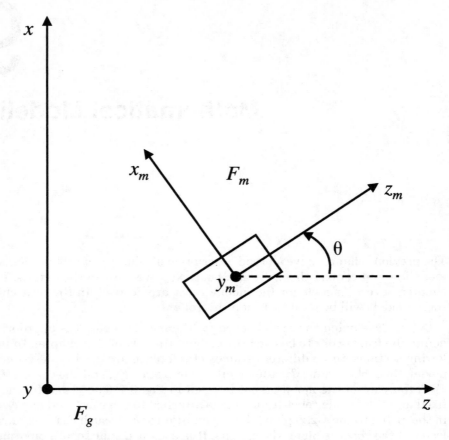

FIGURE 9.1   Top view of the global and mobile robot frames.

Therefore, the vehicle's dynamic properties, such as mass or center of gravity do not enter into the equations. To derive this model, the nonholonomic constraints of the system are utilized.

Nonholonomic constraints

If a system has restrictions in its velocity, but those restrictions do not cause restrictions in its positioning, the system is said to be nonholonomically constrained. Viewed another way, the system's local movement is restricted, but its global movement is not. Mathematically, this means that the velocity constraints cannot be integrated to position constraints. It is easy to understand the concept of nonholonomic constraints with a familiar example: the parallel parking maneuver. When a driver arrives next to a parking space, he cannot simply slide his car sideways into the spot. The car is not capable of sliding side-

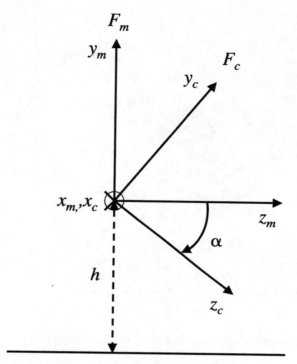

FIGURE 9.2   Side view of the camera frame's location on the mobile robot.

ways and this is the velocity restriction. However, by moving the car forward and backward and turning the wheels, the car can be moved into the parking space. Ignoring the restrictions caused by external objects, the car can be located at any position with any orientation, despite the lack of sideways movement.

The nonholonomic constraints of each wheel of the mobile robot are shown in Fig. 9.3. The wheel's velocity is in the direction of rolling. There is no velocity in the perpendicular direction. This model assumes that there is no wheel slippage.

Unicycle kinematic model

The simplest kinematic model for a mobile robot is given by the unicycle model. The unicycle can be completely characterized by three state variables: z and x give the position and θ gives the orientation. In this model, the y_m-axis goes through the point where the unicycle makes contact with the ground. Applying the nonholonomic constraints to the unicycle gives the following model:

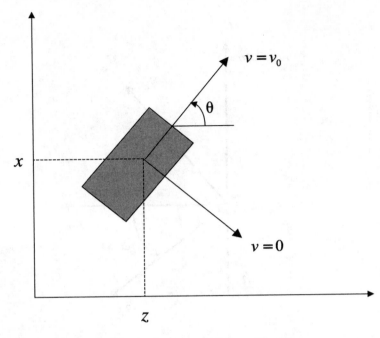

FIGURE 9.3   The velocity constraints on a rolling wheel with no slippage.

$$
\begin{bmatrix} \dot{x} \\ \dot{z} \\ \dot{\theta} \end{bmatrix} = \begin{bmatrix} v \sin \theta \\ v \cos \theta \\ \omega \end{bmatrix}
\tag{9.1}
$$

The inputs to the system are v and ω. The linear velocity, v, is in the direction of the z_m-axis, and ω is the steering input that controls the angular velocity.

Global coordinate model

The exact position and orientation of the car in some global coordinate system can be described by four variables. Fig. 9.4 shows each of the variables. The (z,x) coordinates give the location of the center of the rear axle. The car's angle, with respect to the x-axis, is given by θ. The steering wheels' angle, with respect to the car's longitudinal axis, is given by ϕ.

From the constraints shown in Fig. 9.3, the velocity of the car in the x- and z-directions is given as

$$
\dot{x} = v_1 \sin \theta
\tag{9.2}
$$

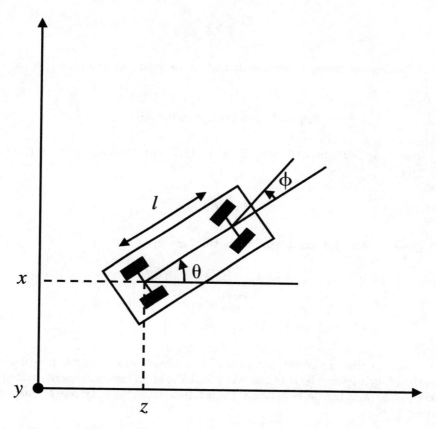

FIGURE 9.4 The global coordinate system for the car.

$$\dot{z} = v_1 \cos \theta \tag{9.3}$$

where v_1 is the linear velocity of the rear wheels. The location of the center of the front axle (z_1, x_1) is given by

$$x_1 = x + l \sin \theta \tag{9.4}$$

$$z_1 = z + l \cos \theta \tag{9.5}$$

and the velocity is given by

$$\dot{x}_1 = \dot{x} + l \, \dot{\theta} \cos \theta \tag{9.6}$$

$$\dot{z}_1 = \dot{z} - l\,\dot{\theta}\sin\theta \qquad (9.7)$$

Applying the no-slippage constraint to the front wheels gives

$$\dot{x}_1\cos(\theta + \phi) = \dot{z}_1\sin(\theta + \phi) \qquad (9.8)$$

Inserting (9.2) and (9.3) into (9.6) and (9.7) and then into (9.8) and solving for $\dot{\theta}$ yields

$$\dot{\theta} = \frac{\tan\phi}{1}\,v_1 \qquad (9.9)$$

The complete kinematic model is then given as

$$\begin{bmatrix} \dot{x} \\ \dot{z} \\ \dot{\theta} \\ \dot{\phi} \end{bmatrix} = \begin{bmatrix} \sin\theta \\ \cos\theta \\ \frac{\tan\phi}{l} \\ 0 \end{bmatrix} v_1 + \begin{bmatrix} 0 \\ 0 \\ 0 \\ 1 \end{bmatrix} v_2 \qquad (9.10)$$

where v_1 is the linear velocity of the rear wheels and v_2 is the angular velocity of the steering wheels. It is assumed that the steering angle is restricted so that $\phi \in (-\frac{\pi}{2}, \frac{\pi}{2})$ and $\tan\phi$ is defined. In practical situations, this is a reasonable assumption.

Path coordinate model

Although the global model is useful for performing simulations, its use for path following is limited in practice. This is because on the hardware implementation, the sensors may not be able to detect the car's location with respect to some global coordinates. On the FLASH car, the sensors can only detect the car's location with respect to the desired path. Therefore, a more useful model is one that describes the car's behavior in terms of the path coordinates.

The path coordinates are shown in Fig. 9.5. The perpendicular distance between the rear axle and the path is given by d. The angle between the car and the tangent to the path is $\theta_p = \theta - \theta_t$. Starting at some arbitrary initial position, the distance traveled along the path is given by s, the arc length.

According to the article, "Control of chained systems: Application to path following and time-varying point-stabilization of mobile robots," (Samson 1995), if it is assumed that the path to be followed is smooth and its curvature, denoted by $c(s)$, is differentiable, then the system can be transformed into the path coordinate model. The curvature is defined as

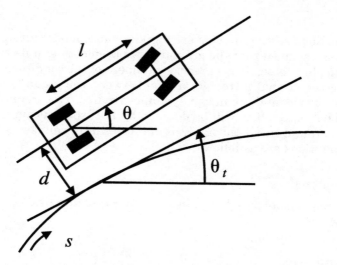

FIGURE 9.5   The path coordinates for the car.

$$c(s) = \frac{d\theta_t}{ds} \tag{9.11}$$

From this definition, $\dot{\theta}_t$ is given to be

$$\dot{\theta}_t = c(s)\dot{s} \tag{9.12}$$

The velocity along the path is

$$\dot{s} = v_1 \cos\theta_p + \dot{\theta}_t d \tag{9.13}$$

and the velocity perpendicular to the path is

$$\dot{d} = v_1 \sin\theta_p \tag{9.14}$$

Combining (9.12), (9.13), (9.14), and the definition of θ_p, the car's kinematic model, in terms of the path coordinates, is given in *Robot Motion Planning and Control*, (De Luca 1999):

$$
\begin{bmatrix} \dot{s} \\ \dot{d} \\ \dot{\theta}_p \\ \dot{\phi} \end{bmatrix} =
\begin{bmatrix} \frac{\cos\theta_p}{1 - dc(s)} \\ \sin\theta_p \\ \frac{\tan\phi}{l} - \frac{c(s)\cos\theta_p}{1 - dc(s)} \\ 0 \end{bmatrix} v_1 +
\begin{bmatrix} 0 \\ 0 \\ 0 \\ 1 \end{bmatrix} v_2 \tag{9.15}
$$

Ground curves

In the previous section, to convert the kinematic model into path coordinate form it was assumed that the ground curve is smooth and its curvature is differentiable. In this section, a precise statement of the conditions for the ground curve is given so that a transformation between the ground and image plane can be developed, and the image dynamics studied. The ground curve that the mobile robot is to follow will be denoted by Γ. If some assumptions are made about the ground curve, the analysis of these curves can be made much easier. The assumptions are as follows:

1. Γ is analytic.
2. Γ can be parameterized by y_c in the camera coordinates.
3. $\frac{\partial \Gamma}{\partial y} = 0$

1. Γ is analytic.

A curve is analytic at a point of Γ if it can be represented by a Taylor series with some radius of convergence. This implies the existence and continuity of all derivatives of Γ.

2. Γ can be parameterized by y_c.

By this assumption, given any y_c, the x_c corresponding to it is unique. If the x_c coordinate were not unique, there would be more than one possible path to follow. Thus, a higher-level decision would need to be made to determine in which direction to proceed.

3. $\frac{\partial \Gamma}{\partial y} = 0$

This is a formal statement of the previous assumption that the robot's environment is such that there is no movement in the y-direction and that the y-axis and y_m-axis remain parallel at all times.

This path coordinate model is still valid using these restrictions. This is because the transformation from global to path coordinates only requires that the path be smooth and its curvature differentiable.

Image dynamics

This section describes the relationship between the ground curve, the curve in the image plane, and the dynamics of the image curve, as given in the paper "Vision Guided Navigation for a Nonholonomic Mobile Robot," (Ma 1999).

By the previous assumptions made about the ground curve, the points of Γ can be parameterized by y_c. Thus, at any time, t, a point on the ground curve, P, can be specified in the camera frame by $(\gamma_x(y_c, t), y_c, \gamma_z(y_c, t))$. An explicit expression can be derived for $\gamma_z(y_c, t)$. From Fig. 9.6, it can be seen that

$$\gamma_z(y_c, t) = \frac{h}{\sin \alpha} + \frac{y_c}{\tan \alpha} = \frac{h + y_c \cos \alpha}{\sin \alpha} \tag{9.16}$$

If y_c is a fixed point in the camera frame, the time evolution of the system can be characterized by $\gamma_x(y_c, t)$ because γ_z is not a function of time.

The ground curve can be represented in the image plane by its orthographic projection, $(\gamma_x(y_c, t), y_c)$, and also by its perspective projection, given by

$$X(y_c, t) = f \frac{\gamma_x}{\gamma_z} \tag{9.17}$$

$$Y(y_c, t) = f \frac{y_c}{\gamma_z} \tag{9.18}$$

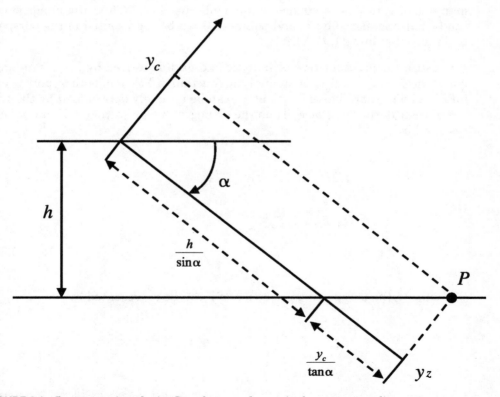

FIGURE 9.6 Representation of point P on the ground curve in the camera coordinates.

where f is the focal length of the camera. Inserting (9.16) into (9.17) and (9.18) gives

$$X(y_c, t) = f \frac{\gamma_x(y_c, t) \sin \alpha}{h + y_c \cos \alpha} \qquad (9.19)$$

$$Y(y_c, t) = f \frac{y_c \sin \alpha}{h + y_c \cos \alpha} \qquad (9.20)$$

At any time t, the derivative of Y with respect to y_c is given by

$$\frac{\partial Y(y_c, t)}{\partial y_c} = \frac{hf \sin\alpha}{(h + y_c \cos\alpha)^2} \qquad (9.21)$$

Because it is assumed that $\alpha \in (0, \pi/2)$, if $y_c \neq h/\cos \alpha$, then Y is a smooth function of y_c and $(\partial Y(y, t))/\partial y_c^c$ is invertible. By the inverse function theorem, the mapping of y_c to Y is one to one in a neighborhood of P. Thus, the image curve can be parameterized by Y and points of Γ can be represented in the perspective projection by $\lambda_X(Y, t), Y$.

Because the ground curve is analytic and parameterized by y_c, γ_x is an analytic function of y_c and λ_X is an analytic function of Y. Then, both γ_x and λ_X are infinitely differentiable and, at any y_c, can be uniquely determined by the values of their derivatives at y_c. Then the orthographic projection of Γ can be represented by

$$\xi_{i+1} = \frac{\partial^i \gamma_x(y_c, t)}{\partial y_c^i} \ (i = 0, 1, \ldots)$$

$$\xi^i \equiv (\xi_1, \xi_2, \ldots, \xi_i)^T$$

$$\xi^i \equiv \xi^\infty \qquad (9.22)$$

Similarly, the perspective projection of Γ can be represented by

$$\zeta_{i+1} = \frac{\partial^i \lambda_X(Y, t)}{\partial Y^i} \ (i = 0, 1, \ldots)$$

$$\zeta^i \equiv (\zeta_1, \zeta_2, \ldots, \zeta_i)^T$$

$$\zeta \equiv \zeta^\infty \qquad (9.23)$$

The two systems of equations (9.22) and (9.23) are linearly related. If one set of coefficients is known, the other can be found through multiplication by a non-singular lower triangular matrix. The complete proof of this statement using mathematical induction is given in (Ma 1999). Thus, ζ_i is a function of $\xi_1, \xi_2, ..., \xi_i$ in general.

The 3×3 matrix relating ζ^3 and ξ^3 can be found as follows. By definitions (9.22) and (9.23)

$$\xi_1 = \gamma_x \tag{9.24}$$

$$\zeta_1 = \lambda_X \tag{9.25}$$

Plugging (9.19) into (9.25) gives

$$\zeta_1 = \frac{f \sin \alpha}{h + y_c \cos \alpha} \xi_1 \tag{9.26}$$

The relation between ξ_2 and ζ_2 can be found by differentiating ζ_1

$$\zeta_2 \frac{\partial Y}{\partial y_c} = \frac{\partial \zeta_1}{\partial y_c} \tag{9.27}$$

Substituting (9.26) into this equation and solving for ζ_2 gives the result in terms of ξ_1 and ξ_2

$$\zeta_2 = -\frac{\cos \alpha}{h} \xi_1 + \frac{h + y_c \cos \alpha}{h} \xi_2 \tag{9.28}$$

This is a linear combination of ξ_1 and ξ_2. Similarly, ζ_3 is determined to be

$$\zeta_3 = \frac{(h + y_c \cos \alpha)^3}{fh^2 \sin \alpha} \xi_3 \tag{9.29}$$

Representing the previous equations in matrix form gives the relationship between ξ^3 and ζ^3

$$\zeta^3 = \begin{bmatrix} \frac{f \sin \alpha}{h + y_c \cos \alpha} & 0 & 0 \\ -\frac{\cos \alpha}{h} & \frac{h + y_c \cos \alpha}{h} & 0 \\ 0 & 0 & \frac{(h + y_c \cos \alpha)^3}{fh^2 \sin \alpha} \end{bmatrix} \xi^3 \tag{9.30}$$

In the implementation, the camera will have a finite focal length, so the measurements from the image will be the ζ coefficients, the perspective projection. However, the control laws and analysis to follow are done using the orthographic projection coefficients, ξ. Because of their equivalence from the previous result, the properties of the orthographic projection system also hold for the perspective projection system.

The change in the arc length of a curve is given by

$$s'(y_c) = \sqrt{\left(\frac{\partial \gamma_x}{\partial y_c}\right)^2 + 1 + \left(\frac{\partial \gamma_z}{\partial y_c}\right)^2} \qquad (9.31)$$

The curvature of a ground curve is defined by (9.11) as

$$c(s) = \frac{d\theta_t}{ds} \qquad (9.32)$$

This result only holds for planar curves. In the global frame, the Γ is planar. However, in the camera frame, Γ moves in all three coordinates. For a curve in $\mathcal{R}^3$, the curvature (parameterized by y_c) is defined as

$$c(y_c) = \left|\frac{dT}{ds}\right| = \frac{|\Gamma'(y_c) \times \Gamma''(y_c)|}{|\Gamma'(y_c)|^3} \qquad (9.33)$$

where T is the tangent vector of Γ at y_c. Because the derivatives of Γ are given by

$$\Gamma'(y_c) = \left(\frac{\partial \gamma_x}{\partial y_c}, 1, \cot \alpha\right)^{\mathrm{T}} \qquad (9.34)$$

$$\Gamma''(y_c) = \left(\frac{\partial^2 \gamma_x}{\partial y_c}, 0, 0\right)^{\mathrm{T}} \qquad (9.35)$$

(9.33) becomes

$$c(y_c) = \frac{\sqrt{\cot^2 \alpha + 1}\left(\frac{\partial^2 \gamma_x}{\partial y_c^2}\right)}{\left(\sqrt{\left(\frac{\partial \gamma_x}{\partial y_c}\right)^2 + 1 + \cot^2 \alpha}\right)^3} \qquad (9.36)$$

Another important result from (Ma 1999) concerns a special case of analytic ground curves, linear curvature curves. By definition of linear curvature, the change in curvature with respect to the arc length is constant but not zero. Let $k = c'(s)$. Because $c'(s) = \frac{c'(\gamma_c)}{s'(\gamma_c)}$, rearranging and solving for $\frac{\partial^3 \gamma_x}{\partial y_c^3}$ results in

$$\frac{\partial^3 \gamma_x}{\partial y_c^3} = \xi_4 = \frac{\dfrac{k(1 + \cot^2\alpha + \xi_2^2)^3}{\sqrt{1 + \cot^2\alpha}} + 3\xi_2\xi_3^2}{1 + \cot^2\alpha + \xi_2^2} \tag{9.37}$$

This result shows that for $i \geq 4$, ξ_i is a function of only ξ_1, ξ_2, and ξ_3. So all of the information for a linear curvature analytic ground curve is captured in ξ^3.

As previously stated, the dynamics of the image curve can be reduced to the dynamics of $\gamma_x(y_c, t)$ because γ_z is only a function of y_c. So, as the mobile robot moves, $(\gamma_x(y_c, t), y_c, \gamma_z(y_c))$ changes with the rotational velocity ω and linear velocity v. The movement of a point in the camera frame due to the robot's linear velocity is given by

$$\begin{bmatrix} \dot\gamma_x \\ \dot y_c \\ \dot\gamma_z \end{bmatrix} = \begin{bmatrix} 0 \\ \sin\alpha \\ \cos\alpha \end{bmatrix} v \tag{9.38}$$

The movement due to the robot's rotational velocity is given by

$$\begin{bmatrix} \dot\gamma_x \\ \dot y_c \\ \dot\gamma_z \end{bmatrix} = - \begin{bmatrix} \gamma_x \\ y_c \\ \gamma_z \end{bmatrix} \times \begin{bmatrix} 0 \\ \omega\cos\alpha \\ -\omega\sin\alpha \end{bmatrix} \tag{9.39}$$

where $(0, \omega\cos\alpha, -\omega\sin\alpha)^{\mathrm{T}}$ is the vector along the axis of rotation with length ω. Thus,

$$\dot\gamma_x = -(y_c\sin\alpha + \gamma_z\cos\alpha)\omega \tag{9.40}$$

$$\dot\gamma_c = -(v_c\sin\alpha - \gamma_x\omega\sin\alpha) \tag{9.41}$$

Using the fact that

$$\dot\gamma_x = \frac{\partial\gamma_x(y_c, t)}{\partial t} + \frac{\partial\gamma_x(y_c, t)}{\partial y_c}\dot y_c \tag{9.42}$$

the dynamics of a point in the image plane are

$$\frac{\partial \gamma_x(y_c, t)}{\partial t} = \left(\frac{\partial \gamma_x}{\partial y_c} \sin \alpha \right) v - \left(\frac{y_c}{\sin \alpha} + h \cot \alpha + \gamma_x \frac{\partial \gamma_x}{\partial y_c} \sin \alpha \right) \omega \qquad (9.43)$$

Thus, $\dot{\xi}_1$ is given by (9.43). Differentiating twice, with respect to y_c, gives the image plane dynamics for a linear curvature curve

$$\dot{\xi}^3 = f_1^3 \omega + f_2^3 v \qquad (9.44)$$

where

$$f_1^3 = - \begin{bmatrix} \xi_1 \xi_2 \sin \alpha + h \cot \alpha + \dfrac{y_c}{\sin \alpha} \\ \xi_1 \xi_3 \sin \alpha + \xi_2^2 \sin \alpha + \dfrac{1}{\sin \alpha} \\ \xi_1 \xi_4 \sin \alpha + 3 \xi_2 \xi_3 \sin \alpha \end{bmatrix}$$

$$f_2^3 = \begin{bmatrix} \xi_2 \sin \alpha \\ \xi_3 \sin \alpha \\ \xi_4 \sin \alpha \end{bmatrix} \qquad (9.45)$$

Dynamic Model

Another model has been developed that takes into account the dynamics of the robotic car. This model gives a more accurate depiction of the car's behavior. The nonholonomic constraints of the system are utilized. This section gives the steps to derive the dynamic model.

For simplicity, the front and rear wheels have been collapsed into a single wheel in the front and a single wheel in the rear. This is known as the bicycle model. Denote (z, x) as being the position of the center of gravity, and θ as the orientation of the vehicle with respect to the z-axis, and ϕ as the steering angle between the front wheel and the body axis. See Fig. 9.7. Let (z_1, x_1) denote the position of the rear axle, and (z_2, x_2) denote the position of the front axle.

This defines

$$z_1 = x - b \cos \theta$$

$$x_1 = y - b \sin \theta$$

$$\dot{z}_1 = x + b\dot{\theta} \sin \theta$$

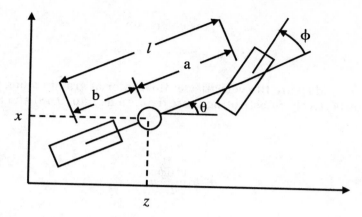

FIGURE 9.7 Parameters for the dynamic model.

$$\dot{x}_1 = y - b\dot{\theta} \cos\theta$$

$$z_2 = z + a \cos\theta$$

$$x_2 = x + a \sin\theta$$

$$\dot{z}_2 = z - a\dot{\theta} \sin\theta$$

$$\dot{x}_2 = x + a\dot{\theta} \cos\theta$$

The nonholonomic constraints are then written for each wheel, resulting in

$$\dot{z}_1 \sin\theta - \dot{x}_1 \cos\theta = 0 \tag{9.46}$$

$$\dot{z}_2 \sin(\theta + \phi) - \dot{x}_2 \cos(\theta + \phi) = 0 \tag{9.47}$$

Substituting the definition of $(\dot{z}_1, \dot{x}_1)$ and $(\dot{z}_2, \dot{x}_2)$ into (9.46) and (9.47) gives

$$\dot{z} \sin\theta - \dot{x} \cos\theta + b\dot{\theta} = 0 \tag{9.48}$$

$$\dot{z} \sin(\theta + \phi) - \dot{x} \cos(\theta + \phi) + a\dot{\theta} \cos\theta = 0 \tag{9.49}$$

Using the coordinate frame of the vehicle, as shown in Fig. 9.8, define the u-axis as being that which is along the length of the vehicle, and the ω-axis normal to the u-axis. Then the velocity of the vehicle in the global frame is given as

$$\dot{z} = v_u \cos \theta - v_w \sin \theta$$

$$\dot{x} = v_u \sin \theta + v_w \cos \theta$$

where v_u and v_w are the velocities of the center of gravity along the u- and w-axes, respectively. Substituting these definitions into (9.48) and (9.49) gives

$$v_w = \dot{\theta} b \qquad (9.50)$$

$$\dot{\theta} = \frac{\tan\phi}{l} v_u \qquad (9.51)$$

Thus, the derivatives of the nonholonomic equations are

$$\dot{v}_w = \ddot{\theta} b \qquad (9.52)$$

$$\ddot{\theta} = \frac{\tan\phi}{l} \dot{v}_u + \frac{v_u}{l \cos^2\phi} \dot{\phi} \qquad (9.53)$$

Now that the kinematic constraints are in a more useful format, the dynamic equations can be derived. These dynamic equations are similar to those found in "Path-Tracking for Car-Like Robots with Single and Double Steering" (Desantis 1995), with two added assumptions. These added assumptions are that there is no friction force between the wheels and the vehicle, and that the rear wheels are locked to be in the same orientation as the vehicle. Other assumptions, used both in (Desantis 1995) and here, are that there is no slip at the wheel, and that the driving force, based on the radius of the wheel and the drive torque, can be modeled as acting at the center of the rear wheels. With the

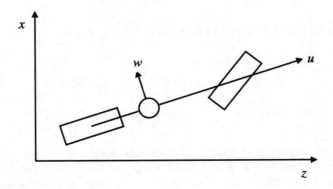

FIGURE 9.8 The vehicle's coordinate frame.

slippage assumption comes a pair of forces, one acting at each wheel, perpendicular to that wheel. The forces involved in this derivation are shown in Fig. 9.9.

The resultant dynamic equations are

$$\dot{v}_u = v_w\dot{\theta} - \frac{F_F \sin \phi}{m} + \frac{F_D}{m} \tag{9.54}$$

$$\dot{v}_w = -v_u\dot{\theta} + \frac{F_F \cos \phi}{m} + \frac{F_R}{m} \tag{9.55}$$

$$\ddot{\theta} = \frac{aF_F \cos \phi}{J} - \frac{bF_R}{J} \tag{9.56}$$

where m is the mass of the vehicle and J is the mass moment of inertia about the center of gravity. F_D is the driving force, which is applied at the rear axle along the u-axis, and F_F and F_R are the resultant lateral forces on the front and rear tires, respectively, as shown in Fig. 9.9.

Solving (9.56) in terms of F_R gives

$$F_R = \frac{aF_F \cos \phi}{b} - \frac{J\ddot{\theta}}{b} \tag{9.57}$$

Substituting this equation with (9.52) into (9.55) gives

$$\ddot{\theta}b = -v_u\dot{\theta} + \frac{F_F \cos \phi}{m} + \frac{aF_F \cos \phi}{bm} - \frac{J\ddot{\theta}}{bm} \tag{9.58}$$

Solving for F_F gives

$$F_F = \frac{b^2m + J}{l \cos \phi}\ddot{\theta} + \frac{bm}{l \cos \phi}v_u\dot{\theta} \tag{9.59}$$

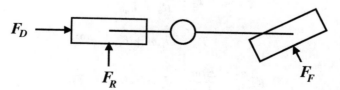

FIGURE 9.9 The forces acting on the vehicle.

Placing (9.50), (9.51), (9.53), and (9.59) into (9.54) gives

$$\dot{v}_u = \frac{v_u(b^2 m + J)\tan\phi}{\gamma}\dot{\phi} + \frac{l^2 \cos^2\phi}{\gamma}F_D \qquad (9.60)$$

$$\gamma = \cos^2\phi[l^2 m + (b^2 m + J)\tan^2\phi] \qquad (9.61)$$

Placing (9.50) and (9.51) into the definitions of velocity and choosing the states to be $X = [z, x, \theta, v_u, F_D, \phi]'$ gives

$$\dot{z} = \left(\cos\theta - \frac{b\tan\phi}{l}\sin\theta\right)v_u$$

$$\dot{x} = \left(\sin\theta + \frac{b\tan\phi}{l}\cos\theta\right)v_u$$

$$\dot{\theta} = \frac{\tan\phi}{l}$$

$$\dot{v}_u = \frac{v_u(b^2 m + J)\tan\phi}{\gamma}\dot{\phi} + \frac{l^2 \cos^2\phi}{\gamma}F_D$$

$$\dot{F}_D = f(F_D, v_u, u_1)$$

$$\dot{\phi} = f(\phi, u_2) \qquad (9.62)$$

Here, γ is defined by (9.61) and u_1 and u_2 are the input voltages, ranging from -5V to 5V. The final step in deriving the dynamics of the system is to find equations for the driving force and the steering servo. Since it is known that the driving force is transmitted to the vehicle through an armature-controlled DC motor, the assumed model is derived as follows:

$$T_m = K_m I_a$$

$$I_a = \frac{u_1 - K_b \omega}{R_a + L_a s}$$

$$T_m = J_m \dot{\omega} + b_m \omega + T_D$$

$$K_m \frac{u_1 - K_b \omega}{R_a + L_a s} = J_m \dot{\omega} + b_m \omega + T_D \qquad (9.63)$$

where K_m, K_b, R_b, L_a, J_m, and b_m are motor constants; ω is the angular velocity of the motor; and T_D is the driving torque. Solving (9.63) for the driving torque and ignoring the higher-order terms, which are three orders of magnitude less or smaller, yields

$$\dot{T}_D = -\frac{R_a}{L_a}T_D - \frac{K_mK_b + R_ab_m}{L_a}\omega + \frac{K_m}{L_a}u_1 \tag{9.64}$$

Solving T_D and ω in terms of $\dot{F}_D$ and v_u gives

$$T_D = \frac{N_mR_\omega}{N_\omega}F_D \tag{9.65}$$

$$\omega = \frac{N_\omega}{N_mR_\omega}v_u \tag{9.66}$$

Differentiating $\dot{F}_D$ gives the following

$$\dot{F}_D = -\frac{R_a}{L_a}F_D - \frac{(K_mK_b + R_ab_m)N_\omega^2}{L_aN_m^2R_\omega^2}v_u + \frac{K_mN_\omega}{L_aN_mR_\omega}u_1 \tag{9.67}$$

where R_w is the radius of the wheel and N_w and N_m are the number of teeth on the gears that connect the axle and motor, respectively. The servo was assumed, and experimentally proven, to be well represented by a linear first-order system of the form

$$\dot{\phi} = \frac{1}{\tau_s}\phi + C_su_2 \tag{9.68}$$

With (9.67) and (9.68), the full state equations can be rewritten as

$$\dot{z} = \left(\cos\theta - \frac{b\tan\phi}{l}\sin\theta\right)v_u$$

$$\dot{x} = \left(\sin\theta + \frac{b\tan\phi}{l}\cos\theta\right)v_u$$

$$\dot{\theta} = \frac{\tan\phi}{l}$$

$$\dot{v}_u = \frac{v_u(b^2m + J)\tan\phi}{\gamma}\dot{\phi} + \frac{l^2\cos^2\phi}{\gamma}F_D$$

$$\dot{\phi} = \frac{1}{\tau_s}\phi + C_s u_2$$

$$\dot{F}_D = -\frac{R_a}{L_a}F_D - \frac{(K_m K_b + R_a b_m)N_\omega^2}{L_a N_m^2 R_\omega^2}v_u + \frac{K_m N\omega}{L_a N_m R_\omega}u_1 \qquad (9.69)$$

with γ defined by (9.61).

Traction Control

Up until this point, we have considered models for cars that assume that there is no wheel slip. This is a reasonable assumption most of the time, except for when there are icy conditions, when we are trying to perform traction control to achieve fastest acceleration, or when we are trying to perform antilock braking. The theory of traction control and how it can be applied is given extensively in (Kachroo, 1993). Here we simply give a brief description of a nonlinear model of the robotic car that includes wheel slip effects in the dynamics. Please refer to (Kachroo, 1993) to obtain more models as well as specific feedback control designs that take wheel slip into account to perform various control tasks with the car, such as

- Traction control (fastest acceleration)
- Antilock braking
- Fastest deceleration
- Adaptive cruise control
- Lane keeping

We can also extend the laws to perform collision avoidance.

Nonlinear vehicle model

The free body diagram of the vehicle is shown in Fig. 9.10. This model has five degrees of freedom: longitudinal and lateral velocities, yaw rate, and rotational velocities for the front and rear wheels. Although this model is described for the acceleration case only, it can be easily modified for the deceleration case. The following description is taken from (Kachroo, 1993).

The lateral components of the forces of the roads on the tires are F_{yf} and F_{yr}, the longitudinal components are F_{xf} and F_{xr}, where the f and r subscripts refer to the front and rear, respectively. The longitudinal wind thrust is F_{wx}, and the lateral wind thrust is F_{wy}. The effects of camber and self-aligning moments are neglected.

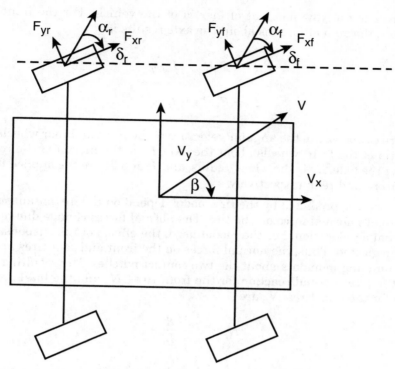

FIGURE 9.10 Schematic diagram of the vehicle model for longitudinal and lateral control.

Summing the lateral forces along the body's y-axis leads to

$$F_{yf} \cos \delta_f + F_{xf} \sin \delta_f + F_{yr} \cos \delta_r + F_{xr} \sin \delta_r + F_{wy} = M_v(\dot{v}_y + v_x r) \qquad (9.70)$$

where M_v is the vehicle's mass, v_x and v_y are the longitudinal and lateral components (on the body axis) of the vehicle velocity, and r is the yaw rate. The angles δ_f and δ_r are the front and rear wheel steering angles.

Summing the longitudinal forces along the body's x-axis gives

$$F_{xf} \cos \delta_f - F_{yf} \sin \delta_f + F_{xr} \cos \delta_r - F_{yr} \sin \delta_r - F_{wx} = M_v(\dot{v}_x - v_y r) \qquad (9.71)$$

The sum of the yaw moments about the center of gravity of the vehicle yields

$$L_f(F_{yf} \cos \delta_f + F_{xf} \sin \delta_f) - L_r(F_{yr} \cos \delta_r + F_{xr} \sin \delta_r) = I\dot{r} \qquad (9.72)$$

where I is the yaw moment of inertia of the vehicle. For the front and rear wheels, the sum of torques about the axle results in

$$T_f - F_{xf}R_w = I_{wf}\dot{\omega}_f \tag{9.73}$$

$$T_r - F_{xr}R_w = I_{wr}\dot{\omega}_r \tag{9.74}$$

where ω_f and ω_r are the angular velocities of the front and rear wheels, I_{wf} is the inertia of the front wheels about the axle, I_{wr} is the inertia of the rear wheels about the axle, R_w is the wheel radius, and T_f and T_r are the applied torques for the front and rear, respectively.

The forces predicted by the tire model depend on the instantaneous value of the road's normal force on the tire. The normal forces change due to the longitudinal acceleration. For the model used, the effects of the suspension system are neglected. Thus, the normal forces on the front and rear tires are obtained by summing moments about the two contact patches. The resulting equations for the total normal reaction for the front tires, N_f, and the total normal reaction for the rear tires, N_r, are

$$N_f = \frac{L_f M_v g}{L_f + L_r} \tag{9.75}$$

$$N_r = \frac{L_r M_v g}{L_f + L_r} \tag{9.76}$$

The nonlinear tire forces are evaluated using the slip angle and the longitudinal slip for each tire. The side slip angle, β, is the angle between the vehicle centerline and the velocity vector of the vehicle's center of gravity. The tire slip angles are

$$\alpha_f = \delta_f - \tan^{-1}\frac{v_y + L_f r}{v_x} \tag{9.77}$$

$$\alpha_r = \delta_r - \tan^{-1}\frac{v_y + L_r r}{v_x} \tag{9.78}$$

The values of the longitudinal slip are

$$\lambda_f = \frac{\omega_f R_w - V_{wf}}{\omega_f R_w} \tag{9.79}$$

$$\lambda_r = \frac{\omega_r R_w - V_{wr}}{\omega_r R_w} \tag{9.80}$$

where V_{wf} and V_{wf} are the longitudinal components of the velocity of the front and rear axles, respectively

$$V_{wf} = v_f \cos\alpha_f \tag{9.81}$$

$$V_{wr} = v_r \cos\alpha_r \tag{9.82}$$

and the magnitudes of the velocities of the front and rear axles, v_f and v_r are

$$v_f = \sqrt{(v_y + L_f r)^2 + v_x^2} \tag{9.83}$$

$$v_r = \sqrt{(v_y + L_r r)^2 + v_x^2} \tag{9.84}$$

When the characteristics of the tire (tire pressure, road and tire surface condition, temperature, etc.) are fixed, the traction and turning forces generated from the tire are solely determined by the tire slip angle and the wheel slip (the tire slip ratio). The longitudinal tire adhesion coefficient is defined as the ratio of longitudinal tire force and the normal force on the same tire. Similarly, the lateral tire adhesion coefficient is defined as the ratio of the lateral tire force and the normal force on the same tire. These adhesion coefficients are nonlinear functions of slip angle and slip ratio. The longitudinal adhesion coefficient versus the slip ratio curve looks like a serpentine curve that gets flatter and flatter for increasing values of the slip angle. On the other hand, the lateral adhesion coefficient versus the slip ratio curve resembles a Gaussian curve that gets flatter for decreasing slip angle values. Typical adhesion coefficients versus slip ratio curves are shown in Fig. 9.11 for various slip angle values. Since adhesion coefficients are functions of slip ratio and slip angle, they can be represented by three-dimensional curves. Fig. 9.12 shows the longitudinal adhesion coefficient on the z-axis, while slip ratio and slip angle are plotted on the x-axis and y-axis, respectively. Fig. 9.13 shows the lateral adhesion coefficient on the z-axis, while slip ratio and slip angle are on the x-axis and y-axis, respectively. These plots were drawn by using approximate analytical functions for the adhesion coefficient versus wheel slip curves. A mathematical serpentine function was used for the longitudinal adhesion coefficient and a Gaussian function was used for lateral adhesion.

This chapter has presented several detailed models for mobile robots, ranging from a simple unicycle kinematic model to more complex models involving vehicle dynamics. Armed with these models, we are now ready to design controllers to achieve our various objectives, such as path following, automatic cruise control, and velocity control.

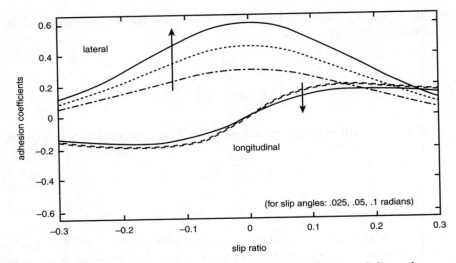

FIGURE 9.11   Longitudinal and lateral adhesion coefficient versus slip ratio and slip angle.

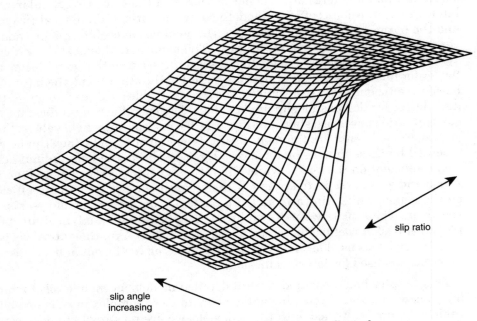

FIGURE 9.12   Longitudinal adhesion coefficient versus slip ratio and slip angle .

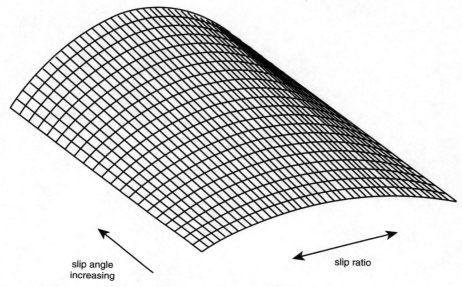

slip angle
increasing

slip ratio

FIGURE 9.13   Lateral adhesion coefficient versus slip ratio and slip angle .

10

Control Design

This chapter uses the models we developed in the previous chapter to design controllers that perform various tasks.

Path Following

As described in the previous chapter, there are three possible tasks that the car could perform: point-to-point stabilization, path following, and trajectory tracking. Point-to-point stabilization requires that the car move from point A to point B with no restrictions on its movement between those two points. With path following, the car must move along a geometric path. Trajectory tracking is similar to path following, except the car must follow a path at a given speed.

In this section, the goal is to design a controller for path following. The car must sense its position with respect to the path and return to the path if its position is off course. The track in the lab contains a white line on a black surface that the car is to follow. In addition, there are magnets beneath the track. The car can sense both of these types of lines, which provide a path for following. A higher-level planner, independent of the controller discussed here, is responsible for determining which type of line to follow. The hardware and software that performs the sensing and planning were discussed in Part 1, "Hardware Implementation."

Path model controller

Before developing the controller for the model given in (9.15), the system is converted into chained form. The $(2,n)$ single-chain form has the following structure (Samson 1995):

$$\dot{x}_1 = u_1$$

$$\dot{x}_2 = u_2$$

$$\dot{x}_3 = x_2 u_1$$

$$\vdots$$

$$\dot{x}_n = x_{n-1} u_1 \tag{10.1}$$

Although the system has two inputs, u_1 and u_2, this model can be considered single input if u_1 is known a priori. For the car model with four states, the (2,4) chained form becomes

$$\dot{x}_1 = u_1$$

$$\dot{x}_2 = u_2$$

$$\dot{x}_3 = x_2 u_1$$

$$\dot{x}_4 = x_3 u_1 \tag{10.2}$$

The states are given as

$$x_1 = s \tag{10.3}$$

$$x_2 = -c'(s)d \tan \theta_p - c(s)(1 - dc(s)) \frac{1 + \sin^2 \theta_p}{\cos^2 \theta_p} + \frac{(1 - dc(s))^2 \tan \phi}{l \cos^3 \theta_p} \tag{10.4}$$

$$x_3 = (1 - dc(s)) \tan \theta_p \tag{10.5}$$

$$x_4 = d \tag{10.6}$$

where the variables are defined in Fig. 9.5, $c(s)$ is the path's curvature, and $c'(s)$ denotes the derivative of c with respect to s. The inputs are defined as follows:

$$v_1 = \frac{1 - dc(s)}{\cos \theta_p} u_1 \tag{10.7}$$

$$v_2 = \alpha_2(u_2 - \alpha_1 u_1) \tag{10.8}$$

where v_1 is the linear velocity of the rear wheels, v_2 is the angular velocity of the steering wheels, and

$$\alpha_1 = \frac{\partial x_2}{\partial s} + \frac{\partial x_2}{\partial d}(1 - dc(s)) \tan \theta_p + \frac{\partial x_2}{\partial \theta_p}\left[\frac{\tan\phi(1 - dc(s))}{l \cos\theta_p} - c(s)\right]$$

$$\alpha_2 = \frac{l \cos^3\theta_p \cos^2\phi}{(1 - dc(s))^2}$$

With the system in chained form, the controller to perform path following can be developed. In this form, path following equates to stabilizing x_2, x_3, x_4 in (10.2) to zero. The input scaling controller from *Robot Motion Planning and Control* (De Luca 1999) is given here.

First, the variables are redefined as follows:

$$\chi = (\chi_1, \chi_2, \chi_3, \chi_4) = (x_1, x_4, x_3, x_2)$$

so the chained form system is then

$$\dot{\chi}_1 = u_1 \tag{10.9}$$

$$\begin{bmatrix} \dot{\chi}_2 \\ \dot{\chi}_3 \\ \dot{\chi}_4 \end{bmatrix} = \begin{bmatrix} 0 & u_1 & 0 \\ 0 & 0 & u_1 \\ 0 & 0 & 0 \end{bmatrix}\begin{bmatrix} \chi_2 \\ \chi_3 \\ \chi_4 \end{bmatrix} + \begin{bmatrix} 0 \\ 0 \\ 1 \end{bmatrix}u_2$$

This system can be transformed into a linear time-invariant system if $u_1(t)$ is bounded and strictly positive (or negative) because such restrictions on u_1 would make χ_1 monotonically increasing with time. So, differentiation with respect to time can be replaced by

$$\frac{d}{dt}(\cdot) = \frac{d\chi_1}{dt}\frac{d}{d\chi_1}(\cdot) = u_1\frac{d}{d\chi_1}(\cdot) \tag{10.10}$$

Dividing by $|u_1|$ gives

$$\frac{1}{|u_1|}\frac{d}{dt}(\cdot) = \text{sign}(u_1)\frac{d}{d\chi_1}(\cdot) \tag{10.11}$$

Using this form of differentiation twice, the system can be represented by

$$\text{sign}(u_1)\frac{d^3\chi_2}{d\chi_1^3} = \text{sign}(u_1)\frac{u_2}{u_1} \tag{10.12}$$

And a stabilizing controller is given by

$$u_2(\chi_2, \chi_3, \chi_4) = -u_1 \text{sign}(u_1)[k_1\chi_2 + k_2\text{sign}(u_1)\chi_3 + k_3\chi_4]$$

As stated in (De Luca 1999), the system (10.9) is controllable if $u_1(t)$ is a "piecewise continuous, bounded, and strictly positive (or negative)" function. With u_1 known, u_2 is left as the only input to the system. The controller for u_2 (with the appropriate restrictions on u_1) becomes

$$u_2 = -k_1|u_1(t)|\chi_2 - k_2 u_1(t)\chi_3 - k_3|u_1(t)|\chi_4 \qquad (10.13)$$

Curvature estimation. The model for the car and the resulting controller given previously require knowledge of the path's curvature. This section describes methods for estimating the path's curvature.

Except for $c(s)$, all of the variables in (9.15) are known or can be measured by sensors on the car. The feedback control algorithm based on this model must know the curvature to calculate the desired inputs v_1 and v_2. The problem is to determine the curvature of the path based on the known or measured variables.

In the FLASH lab, a two-foot-wide track circuit has been built for prototype development. Several constraints have been placed on the path configuration. One is that the path be continuous. Another is that the path be either a straight line or a curve of known constant radius. A sample path showing these constraints is shown in Fig. 10.1. This sample path is made up of straight sections and curves of two different radii. The resulting curvature profile is shown in Fig 10.2.

The curvature, $c(s)$, is defined as

$$c(s) = \frac{d\theta_t}{ds}$$

Therefore, if the path is turning left, $c(s)$ is positive, and if the path is turning right, $c(s)$ is negative. The magnitude of $c(s)$ is $\frac{1}{R}$, where R is the radius of the circle describing the curve.

As a result of the constraints, the curvature of the path as a function of distance is discontinuous and piecewise constant. The derivative of $c(s)$ with respect to distance is zero, except for those locations where the curvature changes. There, the derivative is infinite. Thus, the following assumption is made:

$$c'(s) = 0$$
$$c''(s) = 0$$
$$\vdots$$

with $c'(s)$ denoting the derivative of c with respect to s. The derivatives are taken to be zero with disturbances where the curvature changes.

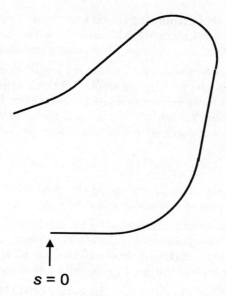

FIGURE 10.1 A sample path showing the constraints.

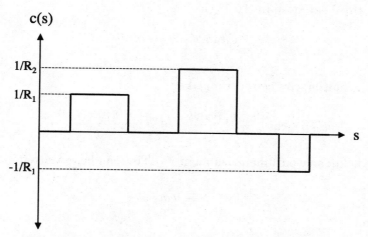

FIGURE 10.2 The curvature of the path in Fig. 10.1 with respect to the path length, s.

The curvature of the path is known; the track is built using pieces with known radii. Therefore, the estimation result can be used to select the actual value. In other words, the calculated curvature need not be used for $c(s)$ in the state equations and controller. Rather, the actual curvature value can be selected based on the outcome of the estimation.

The first method of estimating $c(s)$ is based solely on the steering angle, ϕ. At steady state, the car's steering wheels turn with the curves of the path. This method simply estimates the curvature using the steering wheels' angle.

If the front wheels are fixed at a certain angle, the car will describe a circle of a certain radius. Using (9.10), a MATLAB simulation was used to find the radius, R, described for several values of ϕ. It was found that the relationship between the circle's curvature, $c(s) = 1/R$, and ϕ was nearly a straight line. So the relationship between $c(s)$ and ϕ was approximated to be

$$c(s) = \alpha + \beta\phi \qquad (10.14)$$

where α and β were determined using the method of least squares to fit a line to the data. The sign of $c(s)$ is the same as ϕ.

To make this method more robust to noise, the value of ϕ used in (10.14) can be averaged over several sample periods. By averaging ϕ, this method provides a good estimate even if the car is oscillating about the desired path. However, (10.14) will work only if the car is generally following the desired path.

The second method of estimating the curvature is based on the vehicle's kinematics. If all the variables in (10.15) are known or can be measured, the equation can be solved for $c(s)$.

The third equation in (10.15) is

$$\dot{\theta}_p = \frac{v_1 \tan\phi}{l} - \frac{v_1 c(s)\cos\theta_p}{1 - dc(s)} \qquad (10.15)$$

This equation can be rearranged as

$$c(s)\left[v_1\cos\theta_p + \frac{v_1 d\tan\phi}{l} - \dot{\theta}_p d \right] = \frac{v_1\tan\phi}{l} - \dot{\theta}_p \qquad (10.16)$$

which is linearly parameterizable in $c(s)$. This can be rewritten in the following form:

$$y = wa \qquad (10.17)$$

where

$$y = \frac{v_1\tan\phi}{l} - \dot{\theta}_p \qquad (10.18)$$

$$w = v_1\cos\theta_p + \frac{v_1 d\tan\phi}{l} - \dot{\theta}_p d \qquad (10.19)$$

$$a = c(s) \qquad (10.20)$$

Knowing w and y, a can be obtained using a least squares estimator. We want to find the $\hat{a}$ that minimizes J where

$$J = \int_0^t (y - w\hat{a})^2 dr \tag{10.21}$$

Making $\frac{\partial J}{\partial \hat{a}} = 0$ gives

$$\left[\int_0^t w^2 dr\right]\hat{a} = \int_0^t wy dr \tag{10.22}$$

Differentiating gives an update equation for $\hat{a}$:

$$\dot{\hat{a}} = -Pwe \tag{10.23}$$

where

$$P = \frac{1}{\int_0^t w^2 dr}$$

$$e = w\hat{a} - y$$

and w is defined in (10.19). We can make the equation for P iterative by using the following update equation:

$$\dot{P} = -P^2 w^2 \tag{10.24}$$

where P is initialized to some large value.

The previous estimation methods were simulated using MATLAB. A MATLAB program environment has been created to simulate the car using the kinematic model given in Chapter 9, "Mathematical Modeling." The simulation was run using the controller as given in the section in this chapter titled "Path Model Control." The simulation environment is detailed in Chapter 11, "Simulation Environment."

A path was created in MATLAB to simulate the actual track in the FLASH lab. This path consists of a straight section and a curve of radius 1 m that is followed by another straight section. See Fig. 10.3. The curvature profile is shown in Fig. 10.4. A simulated car was run on the track using the result of the curvature estimate algorithm. Because the curvature of the path was known to be 0 or 1, these actual values were used in the controller. The output of the estimator was utilized to determine which curvature value to use.

First, the curvature estimate based on ϕ was tried. The curvature was calculated using (10.14) with $\alpha = -0.1599$ and $\beta = 4.8975$. For filtering, ϕ was averaged over 10 sample periods. A threshold of 0.5 was used so that if the

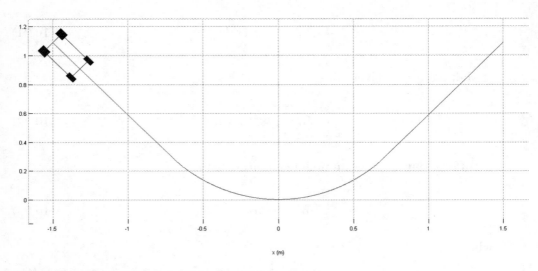

FIGURE 10.3   The path generated using MATLAB.

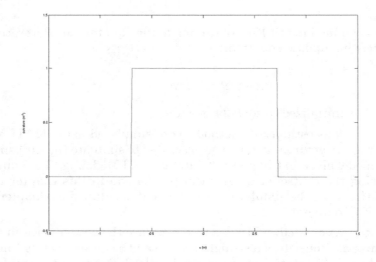

FIGURE 10.4   The curvature profile of the path in Fig. 10.3.

calculated curvature was less than 0.5, a $c(s)$ value of 0 was used. If the calculated curvature was greater than this threshold, $c(s)$ was set to 1.

The car was initially placed so that it was starting on the straight section of the path and oriented so that d and θ_p were both zero. Because of this starting location, there were no transients while the car corrected itself. Fig. 10.5a shows the estimated curvature and the actual curvature plotted together. The actual

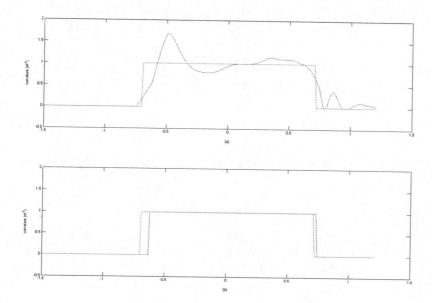

FIGURE 10.5 (*a*) The curvature estimated using only the steering angle, ϕ, with θ_p initially zero. (*b*) The thresholded value.

curvature is shown by a dotted line. Fig. 10.5b shows the thresholded estimate together with the actual curvature. The thresholded value is slightly delayed with respect to the actual curvature.

Next, the car was placed on the path so that θ_p was initially nonzero. This resulted in some transients while the car centered itself on the path. The estimate of the curvature is shown in Fig. 10.6a. The value used for $c(s)$ is shown as the solid line in Fig. 10.6b. Because of the transients, this situation caused $c(s)$ to erroneously have a value of 1 well before the car reached the curve. This method gave a more accurate $c(s)$ during steady state, showing only a slight delay as before.

Next, the curvature estimate based on the kinematic model was simulated. This method used the same initial conditions as the ϕ estimate method.

First, the car was placed on the path so that d and θ_p were initially both zero. The resulting estimate of the curvature is shown in Fig. 10.7a. This estimate was thresholded as before to determine the value for $c(s)$ as 0 or 1. However, to give better performance, hysteresis was used. On the rising edge, the threshold was 0.9; while on the falling edge, the threshold was 0.1. The resulting value for $c(s)$ is shown in Fig. 10.7b. This method seemed to anticipate the curve and thus performed better than the ϕ estimate method.

As with the ϕ estimate method, this method was also tested with a nonzero θ_p. The resulting estimate is shown in Fig. 10.8a. The same hysteresis thresholding

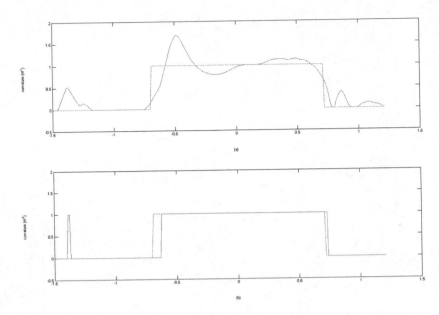

FIGURE 10.6   (*a*) The curvature estimated using only the steering angle, ϕ, with θ_p initially nonzero. (*b*) The thresholded value.

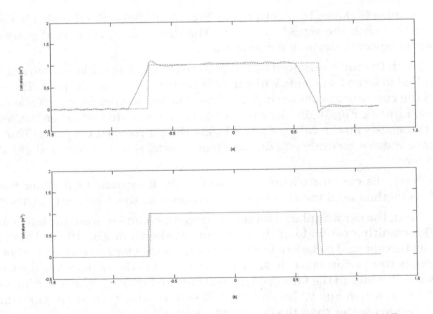

FIGURE 10.7   (*a*) The curvature determined by using the model estimator with θ_p initially zero. (*b*) The thresholded value.

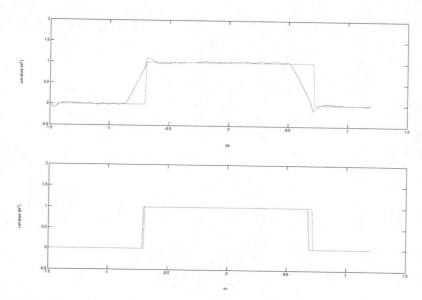

FIGURE 10.8 (*a*) The curvature determined by using the model estimator with θ_p initially nonzero. (*b*) The thresholded value.

was applied in this case and the resulting values for $c(s)$ are shown in Fig. 10.8*b*. This method did not give erroneous results while the car corrected itself on the path.

Another approach was tried with the dynamic curve estimate. After applying the update equation (10.23), $\hat{a}$ was thresholded. If it was greater than 0.5, it was set to 1. If it was less than 0.5, $\hat{a}$ was set to 0. The curvature value for $c(s)$ was then $\hat{a}$. The resulting curvature for both initial conditions is given in Fig. 10.9 and Fig. 10.10. This method performed very well. The estimated curvature matched the actual curvature going from the straightaway to the curve. Coming out of the curve, there was only a slight delay before the estimator determined the correct value for $c(s)$.

Image controller

According to the system given by (9.44), if the robot is tracking the desired curve, then the lateral deviation of the wheel from the curve should be zero and the robot should be moving in the same direction as the ground curve's tangent. Mathematically, this is expressed as

$$\gamma_x(y_c, t)|_{y_c = -h \cos \alpha} = 0$$

$$\frac{\partial \gamma_x(y_c, t)}{\partial y_c}\bigg|_{y_c = -h \cos \alpha} = 0 \qquad (10.25)$$

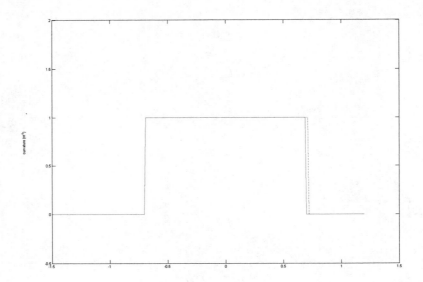

FIGURE 10.9   The curvature determined by thresholding $\hat{a}$ with θ_p initially zero.

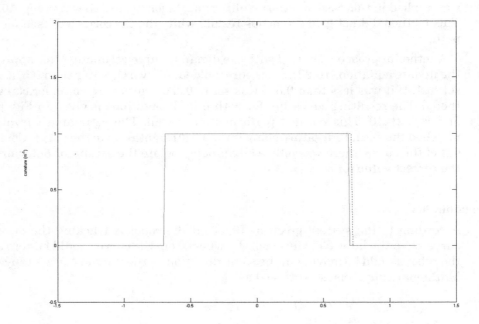

FIGURE 10.10   The curvature determined by thresholding $\hat{a}$ with θ_p initially nonzero.

So the controller should stabilize ξ_1 and ξ_2 to zero.

Combining (10.25) with (9.44) and solving for ω gives the angular velocity when the mobile robot is tracking the desired curve:

$$\omega = (\xi_3 \sin^2\alpha)v \tag{10.26}$$

The control law from (Ma 1999) is given by

$$\omega = \xi_3 \sin^2\alpha\, v_0 + \sin^2\alpha\, \xi_1 v_0 + K_\omega \xi_2 \tag{10.27}$$

$$v = v_0 + \sin^2\alpha\xi_1(\xi_1 + \xi_3)v_0 - K_v\xi_2\text{sign}(\xi_1 + \xi_3)$$

If K_ω and K_v are positive, (10.27) is a stable controller for system (9.44) about $\xi_1 = \xi_2 = 0$, given the desired linear velocity, v_0. This can be shown using the Lyapunov stability theorem on the partial system, $(\xi_1, \xi_2)^T$, with the Lyapunov function defined as

$$V = \xi_1^2 + \xi_2^2 \tag{10.28}$$

Then differentiation gives

$$\dot{V} = -\frac{K_\omega \xi_2^2}{\sin \alpha} - K_\omega \sin \alpha\, \xi_4^2 -$$

$$\xi_2^2 \sin \alpha\, [(K_\omega\xi_1 + \xi_2 \sin^2 \alpha\, v_0) + K_v \,\text{sign}(\xi_1 + \xi_3)](\xi_1 + \xi_3)$$

If ξ_1 and ξ_2 are small enough, then

$$|K_\omega\xi_1 + \xi_2 \sin^2 \alpha\, v_0| \le K_v \tag{10.29}$$

making

$$\dot{V} \le -\frac{K_\omega \xi_2^2}{\sin \alpha} \le 0 \tag{10.30}$$

So, by Lyapunov's stability theorem, the system with controller (10.27) is stable about $\xi_1 = \xi_2 = 0$.

Automatic Cruise Control

Given the framework developed previously for transforming the problem space into arc-length parameters, developing a controller for maintaining a specific distance from objects in front of a vehicle is relatively simple. First, the error term, which must be controlled to maintain a specific distance from objects in front of the vehicle, is defined to be

$$e_h = s_1 - s_2 - h \tag{10.31}$$

Where e_h is the error in the headway, $s_1 - s_2$ is the distance between the two vehicles in terms of the arc-length parameter, and h is the desired headway distance to maintain between the two vehicles. The dynamics of this equation are then given by

$$\dot{e}_h = \dot{s}_1 - \dot{s}_2 = \omega(t) - f(c)v_1 \tag{10.32}$$

where $\omega(t)$ is the estimated velocity of the lead vehicle in terms of arc length at the current sample time, and $f(c)$ represents the dynamics of the follower given by the transformation from the kinematic model in terms of arc length:

$$f(c) = \frac{\cos\theta_p}{1 - dc(s)} \tag{10.33}$$

Clearly, f is a function of the curvature of the path to be followed, and thus the lateral controller is decoupled from the headway controller, as mentioned previously. Also, v_1 is the driving velocity of the input and can be chosen to linearize this system, as the following input:

$$v_1 = \frac{1}{f(c)}[\omega(t) + ke_h] \tag{10.34}$$

When modeled theoretically, $\omega(t)$ is simply $s_1 - s_2$. However, it is obvious that when applied practically, only an estimation of the dynamics of the leading vehicle can be obtained. In this case, the estimate of the distance between the two vehicles is obtained from the headway sensor and is represented by $\hat{\omega}(t)$. The difference in terms of arc length between the two vehicles can be estimated simply from the formula for arc length given by

$$L = \int_a^b \sqrt{\left(\frac{dx}{dt}\right)^2 + \left(\frac{dy}{dt}\right)^2} \tag{10.35}$$

The controller implemented uses the trapezoidal rule to approximate the integral. Given a sampling interval, the difference in arc length between the two vehicles can be found simply by approximating the velocity as the distance between the two vehicles divided by the sampling time. Knowing this value for the current sampling time and for the previous sampling time, the amount of arc length between the two vehicles is estimated by

$$diff_s = \frac{T}{2}\left(\sqrt{\left(\frac{d\hat{x}}{T}\right)^2 + \left(\frac{d\hat{y}}{T}\right)^2} + \sqrt{\left(\frac{d\hat{x}_p}{T}\right)^2 + \left(\frac{d\hat{y}_p}{T}\right)^2}\right)$$ (10.36)

where T is the known sampling time, and $d\hat{x}$, $d\hat{x}_p$, $d\hat{y}$, and $d\hat{y}_p$ are the estimates in the difference between the positions of the two vehicles given by the headway sensor for two consecutive readings. Thus, we see that the velocity of the follower vehicle is given by

$$v_1 = \frac{(1 - dc(s))(\frac{ds_1}{dt} + ke_h)}{\cos\hat{\theta}_p}$$ (10.37)

$$e_h = diff_s - h$$ (10.38)

In this application, $\frac{ds_1}{dt}$ is the estimated velocity of the leader vehicle, given by using two consecutive readings from the headway sensor to estimate the rate of closing, and taking into account the known velocity of the follower vehicle. Thus, as the vehicles move around the path, assuming that the follower vehicle initially has a velocity greater than the leader vehicle, the error term, e_h, will go to zero, and the input velocity of the follower vehicle, v_1, will go to the input velocity of the leader vehicle over time.

Practically speaking, it is clear that under operating conditions, the follower vehicle would have a maximum velocity under which it is operating, as in cruise control. Thus, the follower vehicle will match the speed of the leader vehicle, maintaining the correct headway distance until the leader vehicle's time exceeds the follower vehicle's maximum speed. Another way to see this is the situation where there is no vehicle in front of the vehicle being controlled. In this case, there would be no change in the velocity of the follower vehicle, as there are no obstacles to impede the vehicle's velocity.

Also, it is obvious that you would not want your vehicle to be thrown into reverse should an obstacle suddenly appear within the headway distance, and so the velocity has a lower limit of $0 m/s$. Thus, if for any reason an obstacle appears within the headway distance, the vehicle will simply stop as quickly as possible and allow enough headway distance between the two objects to accumulate before matching the obstacle's velocity.

Actuator Control

This section discusses the lateral and velocity controllers.

Lateral controller

The goal of the lateral controller is to put $\dot{\phi}$ in the form that appears in (De Luca 1999). This form is

$$\dot{\phi} = v_2 \tag{10.39}$$

With this in mind, the first transform of variables becomes

$$u_2 = \frac{v_2 - \frac{1}{\tau_s}\phi}{c_s} \tag{10.40}$$

From here, the control law from (De Luca 1999) is reformulated. The first step in the controller is to convert the kinematic states from global coordinates to path coordinates. Since the velocity, force, and steering equations are not dependent upon the z and x positions of the vehicle, those coordinates can be redefined as being the rear axle of the vehicle. As a result, b in the first two state equations becomes zero. Ignoring the velocity and force equations, the remaining states are

$$\begin{bmatrix} \dot{x} \\ \dot{z} \\ \dot{\theta} \\ \dot{\phi} \end{bmatrix} = \begin{bmatrix} \sin\theta \\ \cos\theta \\ \frac{\tan\phi}{l} \\ 0 \end{bmatrix} v_u + \begin{bmatrix} 0 \\ 0 \\ 0 \\ 1 \end{bmatrix} v_2 \tag{10.41}$$

The path coordinate model then becomes

$$\begin{bmatrix} \dot{s} \\ \dot{d} \\ \dot{\theta}_p \\ \dot{\phi} \end{bmatrix} = \begin{bmatrix} \frac{\cos\theta_p}{1 - dc(s)} \\ \sin\theta_p \\ \frac{\tan\phi}{l} - \frac{c(s)\cos\theta_p}{1 - dc(s)} \\ 0 \end{bmatrix} v_u + \begin{bmatrix} 0 \\ 0 \\ 0 \\ 1 \end{bmatrix} v_2 \tag{10.42}$$

Note that the previous two systems are similar to (9.10) and (9.15), except v_1 has been replaced by v_2. This is the general (2,4) chained form as previously defined in this chapter. With a slight modification to avoid redundant variables, the formula becomes

$$\dot{x}_1 = u_s$$
$$\dot{x}_2 = \omega_2$$
$$\dot{x}_3 = x_2 u_s$$
$$\dot{x}_4 = x_3 u_s \tag{10.43}$$

This chained form is generated using the same variable transformation as before, except that v_1 is replaced by v_u, u_1 by v_s, and u_2 by ω_2. The lateral controller is then given by

$$\omega_2 = -k_1|v_s|\chi_2 - k_2 v_s\chi_3 - k_3|v_s|\chi_4 \tag{10.44}$$

The gains, k_1, k_2, and k_3, are chosen for stability. The main difference between the control law designed here and the one designed for the path model controller is that this one takes the dynamics of the servo into account. This small addition makes for much less error.

Velocity controller

The major shortfall of the lateral controller given in this chapter so far, as well as other path-following control algorithms, is that they assume the velocity is directly controlled. Unfortunately, as the state equations in (9.69) indicate, this is not the case for most systems. In an attempt to keep the advantages of nonlinear control, feedback linearization did not seem to be a viable solution. In that light, a controller has been designed using standard backstepping techniques, as laid out in *Nonlinear Systems* (Khalil 1996), with one small variation.

First, focus on the lower three states of (9.69):

$$\dot{v}_u = \frac{v_u(b^2m + J)\tan\phi}{\gamma}\dot{\phi} + \frac{l^2\cos^2\phi}{\gamma}F_D$$

$$\dot{\phi} = \frac{1}{\tau_s}\phi + C_s u_2$$

$$\dot{F}_D = -\frac{R_a}{L_a}F_D - \frac{(K_mK_b + R_ab_m)N_\omega^2}{L_aN_m^2R_\omega^2}v_u + \frac{K_mN\omega}{L_aN_mR_\omega}u_1 \tag{10.45}$$

Substituting (10.40) into (10.45), and choosing

$$u_1 = \frac{R_\omega N_m L_a}{N_\omega K_m}\left(\omega_1 + \frac{(R_ab_m + K_mK_b)N_\omega^2}{N_m^2R_\omega^2} + \frac{R_a}{L_a}F_D\right) \tag{10.46}$$

results in

$$\dot{v}_u = -\frac{v_u(b^2m + J)\tan\phi}{\gamma}v_2 + \frac{l^2\cos^2\phi}{\gamma}F_D$$

$$\dot{F}_D = \omega_1$$

$$\dot{\phi} = v_2 \tag{10.47}$$

where v_2 is the transformed input variable. Now, define η as

$$\eta = \frac{\gamma}{l^2 \cos^2\phi}\left(-k_{v_1}v_u + \frac{v_u \tan\phi(b^2m + J)}{\gamma}v_2\right) \tag{10.48}$$

and choose

$$V(v_u) = \frac{1}{2}v_u^2 \rightarrow \dot{V} = v_u\dot{v}_u = -k_{v1}v_u^2$$

As a result, the velocity and force equations become

$$\dot{v}_u = \frac{v_u(b^2m + J)\tan\phi}{\gamma}v_2 + \frac{L^2\cos^2\phi}{\gamma}\eta + \frac{l^2\cos^2\phi}{\gamma}\mu$$

$$\mu = F_D - \eta$$

$$\dot{\mu} = \omega_1 - \dot{\eta} \tag{10.49}$$

Following the procedure for backstepping, set

$$\omega_1 = v_1 + \dot{\eta}$$

$$v_1 = -\frac{\partial V}{\partial v_u}g(v_u) - k_{v_2}\mu \rightarrow v_1 = -\frac{v_u l^2\cos^2\phi}{\gamma} - k_{v_2}(F_D - \eta) \tag{10.50}$$

Here η is a function of all the states, including those not within the derivation of the controller. As a result, a more general form of backstepping must be used. Instead of using

$$\omega_1 = v_1 + \frac{\partial\eta}{\partial x}\dot{x}$$

the transform becomes

$$\omega_1 = v_1 + \frac{d\eta}{dt}$$

where

$$\frac{d\eta}{dt} = \frac{d}{dt}\left[\frac{\gamma}{l^2\cos^2\phi}\left(-k_{v1}v_u + \frac{v^u\tan\phi(b^2m + J)}{\gamma}v^2\right)\right]$$

This controller does have one flaw. It drives the velocity to zero. Since the goal of this controller is to drive the velocity to a specific value, a slight change must be made to the states. Specifically, η is changed to

$$\eta = \frac{\gamma}{l^2 \cos^2}\left(-k_{v_1}(v_u - v_d) + \dot{v}_d + \frac{v_u \tan\phi(b^2m + J)}{\gamma}v_2 \right) \qquad (10.51)$$

where v_d, the desired velocity, and its derivative are given by the user upon startup. This leads to the following changes in states:

$$\dot{v}_u = \dot{v}_d - k_{v_1}(v_u - v_d) \rightarrow \dot{e}_v = -k_{v_1}e_v$$

$$e_v = v_u - v_d$$

The state equations then become

$$\dot{s} = \frac{\cos\theta_p}{1 - dc(s)}(e_v + v_d)$$

$$\dot{d} = \sin\theta_p(e_v + v_d)$$

$$\dot{\theta}_p = \left(\frac{\tan\phi}{l} - \frac{c(s)\cos\phi}{1 - dc(s)} \right)(e_v + v_d)$$

$$\dot{e}_v = -\frac{v_u(b^2m + J)\tan\phi}{\gamma}v_2 + \frac{l^2\cos^2\phi}{\gamma}\eta + \frac{l^2\cos^2\phi}{\gamma}\mu$$

$$\dot{\mu} = v_1$$

$$\dot{\phi} = v_2 \qquad (10.52)$$

$$\eta = \frac{\gamma}{l^2\cos^2\phi}\left(-k_{v_1}(v_u - v_d) + \dot{v}_d + \frac{v_u\tan\phi(b^2m + J)}{\gamma}v_2 \right)$$

$$\gamma = \cos^2\phi[l^2m + (b^2m + J)\tan^2\phi] \qquad (10.53)$$

Traction Control

Until now, all the controllers we have built were based on models of car behavior that assumed that there is no wheel slip. This assumption is valid for many cases. However, there are times that we need to build controllers that specifically use the relationship of wheel slip to the car dynamics. As an example, the relationship of wheel slip to braking force is crucial for the effective use of

Antilock Braking Systems (ABS). Similarly, in order to obtain the maximum acceleration possible, the amount of wheel slip present controls the force that drives the vehicle forward. Hence, for designing controllers for ABS or traction control systems in general, we need to use models that incorporate wheel slip dependencies.

Chapter 9 gave a dynamic model of the car that uses the relationship of the adhesion coefficient to the wheel slip. This relationship is the key factor that controls the vehicle behavior using wheel slip. By controlling the amount of wheel slip, we can perform various car control tasks such as

- Antilock braking and antispin acceleration
- Maximum acceleration or deceleration
- "Safe" lateral and longitudinal control in icy conditions

Detailed control designs for these different objectives are given in (Kachroo 1993).

Before implementing the controllers from this chapter on the car itself, it is useful to first simulate them to verify that they work as intended. The next chapter describes a simulation environment for the car and gives simulation results for the path-following controller.

Simulation Environment

This chapter describes a simulation environment in which the control algorithms can be tested. Why simulate the performance of the car rather than immediately program the controller onto the car? One obvious answer is that the simulation provides a way to test the controller without incurring any damage to the car. If the controller causes the car to crash in simulation, then almost certainly it won't work on the actual car.

However, it must be understood that the simulation is just that, a simulation. It is based around a model of the car and so it does not capture all the aspects of the car. Even if the controller works in simulation, that is not proof that it will work when programmed onto the car. If the controller is not robust enough, it may be susceptible to unmodeled dynamics or modeling uncertainty that would cause the car to crash. So the simulation must be understood to be only one step in the process of implementing and testing a control design.

The simulation was written using MATLAB. Produced by the MathWorks Inc., MATLAB is a popular and powerful language for technical computing. It originates from matrix computational software. In fact, matrices are the building blocks for language itself. Variables are stored as matrices whose dimensions need not be defined ahead of time and can be changed during runtime.

MATLAB is now a very powerful language that can be used for algorithm development, data analysis, modeling, and simulation. It has a graphical system that can be used to display data as well as create *graphical user interfaces* (GUIs) for applications. In addition, several "toolboxes" contain specialized functions. These are available in areas such as signal processing, control, and neural networks, and they have files that are specific to those areas.

Included with MATLAB is a program called Simulink. Simulink allows the user to model, simulate, and analyze nonlinear dynamic systems. It has a GUI so the user can drag and drop components into a workspace to create a system model. This means that you are making a block diagram of the system that can be invoked from MATLAB rather than describing the system in some specialized language. Simulink was used in this simulation to describe the (very nonlinear) kinematic model for the car.

Simulation Overview

The rest of this chapter describes the MATLAB simulation environment used for developing and testing the control algorithms used on the FLASH vehicle. This simulation provides, as closely as possible, a program environment similar to that used by the FLASH vehicle. The car's methods of measurement and calculation are the same in the simulation as in the hardware. However, in the simulation, an ideal path is created for the car to follow, the car's movement is given by the kinematic model, and the car's movement is shown using the MATLAB animation toolbox.

A flowchart of the program is shown in Fig. 11.1. First, the initialization involves creating the car and path for animation and placing the car on the path. Next, the car's position on the path is determined and the values needed by the controller are calculated. With these values known, the controller then calculates the necessary velocity and steering inputs to make the car follow the path. These inputs are used in the kinematic model to update the car's position and the animation is then updated to show the car's new location. These steps are repeated until the end of the simulation is reached.

The Simulation Program

This section gives the details of how each step is performed.

Path creation

As previously stated, there are several constraints on the construction of the track. Because of these constraints, the path is assumed to be continuous and the curvature is assumed to be piecewise constant. In addition, it is known that the track is made up of straight sections and curves of constant radius. The simulation has been set up to create a path that contains a straight segment, followed by a curve, followed by another straight segment. The path is defined in the (x,y) global coordinates and its length and the radius of curvature can be defined by the user in the initialization file. The path can be defined using the following equation:

$$
y = \begin{cases} -x + r(1 - \frac{2}{\sqrt{2}}), & x < -\frac{r}{\sqrt{2}} \\ -\sqrt{r^2 - x^2} + r, & -\frac{r}{\sqrt{2}} \le x \le \frac{r}{\sqrt{2}} \\ x + r(1 - \frac{2}{\sqrt{2}}), & x > \frac{r}{\sqrt{2}} \end{cases}
\tag{11.1}
$$

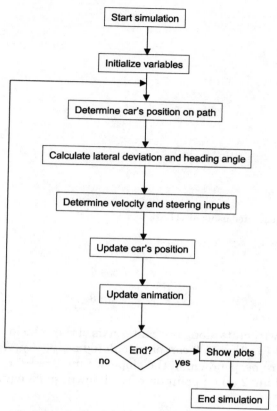

FIGURE 11.1 Flowchart for the MATLAB simulation program.

Here, r is the radius of the curved section of the path. Using $r = 1$ creates the path shown in Fig. 11.2.

Error calculation

With the car on the path, the controller must know where the car is located and how it is oriented. On the FLASH car, there are sensors on the front and rear that detect the presence of the line beneath the car. In the simulation, the distance between the path and the car is found. Then this value is converted to the same representation as on the actual vehicle. Finally, the sensor data is converted to an actual distance.

Calculating the actual error. The vehicle's position is known, (x_0, y_0), as well as its orientation θ and steering angle ϕ. From this, the position of the front sensor can be found as follows.

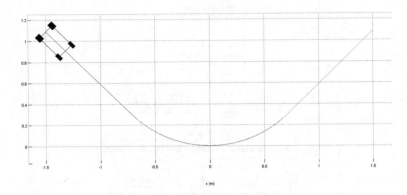

FIGURE 11.2   The path generated using MATLAB.

$$x_1 = x_0 + l \cos \theta \tag{11.2}$$

$$y_1 = y_0 + l \sin \theta \tag{11.3}$$

Knowing two points along the center axis of the vehicle, (x_0, y_0) and (x_1, y_1), the slope of Line 1 in Fig. 11.3 can be calculated as $(y_1 - y_0)/(x_1 - x_0)$. Since Line 1 and Line 2 are perpendicular, the slope of Line 2 is $[-(x_1 - x_0)]/(y_1 - y_0)$. Now the slope of Line 2 and a point on it are known, so its equation is

$$y = -m(x - x_1) + y_1 \tag{11.4}$$

where $m = (x_1 - x_0)/(y_1 - y_0)$.

Next, the point (x_2, y_2) must be determined by finding the intercept of Line 2 and the path. Setting the right side of (11.1) equal to the right side of (11.2) yields the following.

For $x_1 < \frac{-r}{\sqrt{2}}$,

$$x_2 = \frac{mx_1 + y_1 - r(1 - \frac{2}{\sqrt{2}})}{m - 1} \tag{11.5}$$

$$y_2 = -x_2 + r\left(1 - \frac{2}{\sqrt{2}}\right) \tag{11.6}$$

For $\frac{-r}{\sqrt{2}} \le x_1 \le \frac{r}{\sqrt{2}}$,

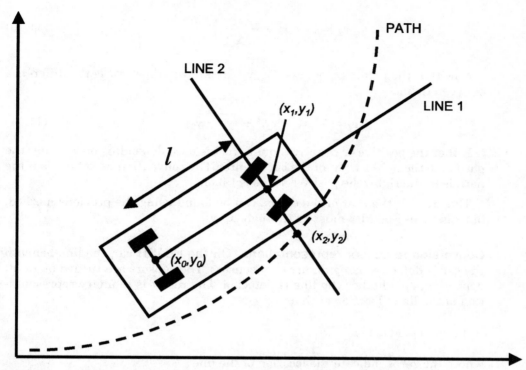

FIGURE 11.3 Errors of the path-following vehicle.

$$x_2 = \frac{-b^2 \pm \sqrt{b^2 - 4ac}}{2a} \qquad (11.7)$$

$$y_2 = -\sqrt{r^2 - x_2^2} + r \qquad (11.8)$$

where the sign of the square root in (11.7) is the same as the sign of m and

$$a = m^2 + 1$$
$$b = -2m^2 x_1 - 2my_1 + 2mr$$
$$c = 2mx_1 y_1 - 2mx_1 r + y_1^2 - 2y_1 r$$

For $x_1 > \frac{r}{\sqrt{2}}$,

$$x_2 = \frac{mx_1 + y_1 - r(1 - \frac{2}{\sqrt{2}})}{m - 1} \qquad (11.9)$$

$$y_2 = x_2 + r\left(1 - \frac{2}{\sqrt{2}}\right) \qquad (11.10)$$

Now that the points (x_1, y_1) and (x_2, y_2) are known, the error is the difference between them.

$$e_f = \sqrt{(x_1 - x_2)^2 + (y_1 - y_2)^2} \qquad (11.11)$$

Either the positive or negative square root is used depending on whether the path is to the left or right of the car's center. The convention used here is if the path is to the right, the positive value is taken.

The error at the rear of the car, e_b, can be found using the previous method, but Line 2 in Fig. 11.3 must go through (x_0, y_0).

Conversion to sensor representation.   On the FLASH car, the line beneath the car is detected using an array of sensors. The sensors are turned on or off depending on whether the line is detected. The result is a binary representation of the line's location such as

```
1110011111111111
```

where the zeros indicate the location of the line.

The error distance calculated in the previous section must be converted to this binary representation.

```
array = ones(s,1);

for k = -0.5*s:0.5*s-1
    if (d > = k*sp) & (d <= (k+1)*sp)
        array(s/2-k) = 0;
    end
end
```

Here, s is the number of sensors, d is the error distance as calculated in the previous section, and sp is the spacing between the sensors. If the line is outside the field of vision for the sensors, the array is all ones.

Conversion to distance.   The binary representation must now be converted back into an actual distance for use by the controller. The following code performs the conversion:

```
error = 0;
num = 0;
val = (s+1)/2;
```

```
for k = 1:s
   if array(k) == 0
      num = num+1;
      error = error+(val-k)*sp;
   end
end

if num ~= 0
   error = error/num;
else
   error = w/2*sign(p);
end
```

where s is the number of sensors, sp is the spacing between the sensors, w is the width of the sensor array, and p is the previously calculated error value. If the array is all ones, the line is outside the range of the sensors and the error is saturated to its maximum value. The sign of the error is then assumed to be the same as p.

Because the rear sensor array is placed directly below the rear axle, the error obtained from that sensor is taken to be d, the car's lateral displacement from the path.

Heading angle calculation

The controller must know the heading angle, θ_p, which is the angle between the car and the path. The value can be calculated using the displacement errors at the front and rear of the car as previously determined, e_f and e_b, and the distance between them, l. Assuming the path directly underneath the car is straight, the heading angle is

$$\theta_p = \tan^{-1}\left(\frac{e_f - e_b}{l}\right) \tag{11.12}$$

In this equation, either the actual errors or the discretized errors can be used. However, to simulate the actual car, the discretized errors should be used.

Control input calculation

The next step in the program is to determine the steering and velocity inputs to move the car along the path. The controller must know the values for d, θ_p, and c. The values for d and θ_p are known from the previous sections. The simulation has been set up to implement the curvature estimation methods.

With these values known, the states x_2, x_3, and x_4 can be calculated using (10.4) through (10.6). The controller is given by (10.1.1) and its output is transformed into steering and velocity inputs by (10.7) and (10.8). The most challenging aspect of the controller implementation is typing in the equations without errors.

Car model

Next, the movement of the car is determined over the sampling period, T. The kinematic model given by (9.10) was implemented in Simulink and is shown in Fig. 11.4. The model uses the current position (x, y, θ, ϕ) as the initial conditions and integrates to determine the car's new position after the inputs are applied for time T. It is assumed that the inputs are constant over the sampling time, as they are on the actual car.

Animation

Finally, the movement is animated to provide a means of viewing the car's behavior. The animation toolbox for MATLAB was used for this purpose. The animation toolbox utilizes Handle Graphics, MATLAB's object-oriented graphics system. This toolbox allows for the animation of any object created in MATLAB. Complete details of this toolbox can be found in the *MATLAB Handbook* (Redfern 2001).

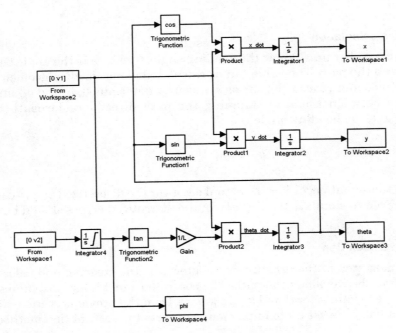

FIGURE 11.4 The Simulink representation of the car's kinematic model.

The first step in using animation is to create the car. The car is simply a rectangle with four wheels attached. The car body is defined using the *patch* command with the appropriate vertices. The wheels are defined as cylinders of necessary height and radius and rotated 90 degrees. Once the individual pieces of the car are created, they must be placed in the proper orientation using the *locate* and *rotate* commands. Finally, they are joined together using the *attach* command so that the entire car can be moved as one piece. In addition, each of the components can be moved individually.

Now, with the car fully defined, it can be located and oriented anywhere on a MATLAB plot. So, given the car's position from the kinematic model, the updated position is obtained by using these commands:

```
locate(car,[x0,y0 0]);
turn(car,'z',(theta0-theta0_prev)*180/pi);
turn(car.tire_fl,'z',(phi0-phi0_prev)*180/pi);
turn(car.tire_fr,'z',(phi0-phi0_prev)*180/pi);
```

This code positions the car at (x_0, y_0) with an orientation of θ_0 and turns the front wheels by an additional ϕ_0.

Simulation Results

This section provides simulation results for the path model controller in varying conditions. The performance of the controllers is discussed and comparisons between them are made. The path used is the same as shown in Fig. 11.2. This controller was tested using both the actual and the discretized errors.

The path model controller discussed in Chapter 10, "Control Design," was implemented using three different forms. The first form used the actual curvature of the path. The second used the ϕ estimation method. The last used the model estimator. The performance of the estimation methods was discussed with respect to their accuracy in determining the curvature. In this section, the performance of the controller using these methods is discussed.

Control using the actual curvature

First, the actual error distances are given in (11.11) and the actual curvature is used in the controller. This is possible because the path was created using (11.1) and the car's location is known. In addition, it is known in which direction along the path the car is traveling. Therefore, the curvature can be determined to be $\pm\frac{1}{R}$ or 0. Figs. 11.5 through 11.7 show the results of applying this controller. The car's path speed, u_1, was held constant at 1.5 m/s. The gains used were $k_1 = \lambda^3$, $k_2 = 3\lambda^2$, and $k_3 = 3\lambda$ with $\lambda = 8$.

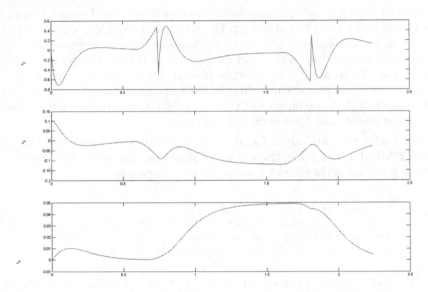

FIGURE 11.5 The states x_2, x_3, and x_4 resulting from using the actual errors and curvature.

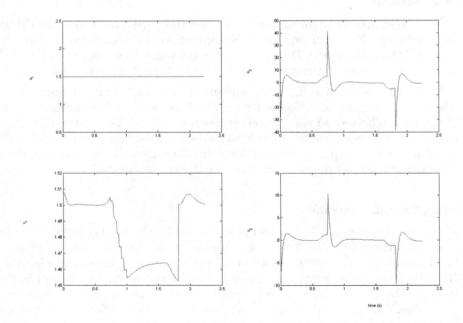

FIGURE 11.6 The control inputs v_1 and v_2 resulting from using the actual errors and curvature.

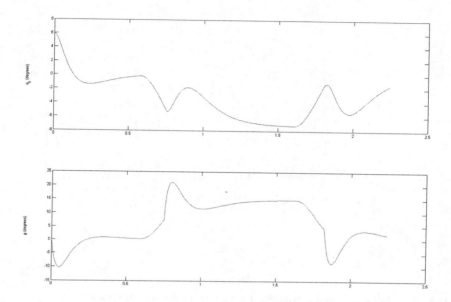

FIGURE 11.7 The heading angle θ_p and steering angle ϕ resulting from using the actual errors and curvature.

There are two important things to notice about the performance of this controller. The first is that even though u_1 is constant, v_1 does not remain constant. u_1 is transformed into v_1 by taking into account the car's state and also the curvature. The result is that the car slows down in the curve.

The second thing to notice is that there are spikes in the steering control input, v_2. These result from the spikes that are present in x_2. These spikes occur exactly where the path changes curvature. At these points, the derivative of the curvature is infinite. However, in the implementation, the derivatives of curvature are set to 0. The discrepancy is seen here as a disturbance in the system.

Next, the same controller was used with the discretized errors. It was assumed that there were 12 sensors spaced 0.2 inches apart. The same gains and initial conditions were used as above. Figs. 11.8 through 11.10 show the results of discretization. Again, the actual curvature is used.

As is to be expected, using the discretized errors caused the control input, and thus the steering angle ϕ, to become much choppier. This resulted in a less smooth trajectory being traversed by the car.

Control using the ϕ estimator

Next, the simulation was run using the ϕ estimator. The results of this algorithm are shown in Figs. 11.11 through 11.13.

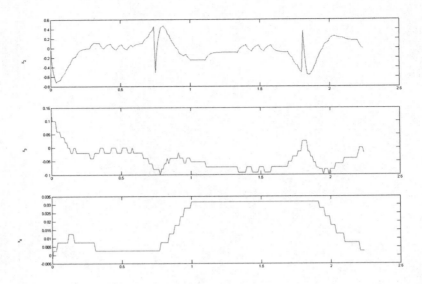

FIGURE 11.8   The states x_2, x_3, and x_4 resulting from using the discretized errors.

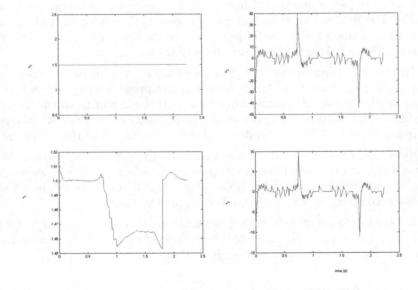

FIGURE 11.9   The control inputs v_1 and v_2 resulting from the discretized errors.

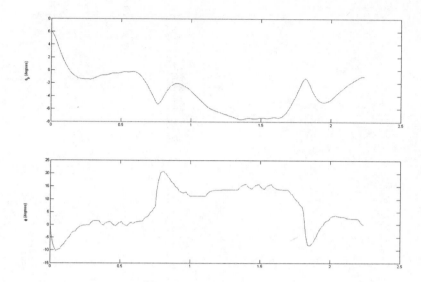

FIGURE 11.10 The heading angle θ_p and the steering angle ϕ resulting from using the discretized errors.

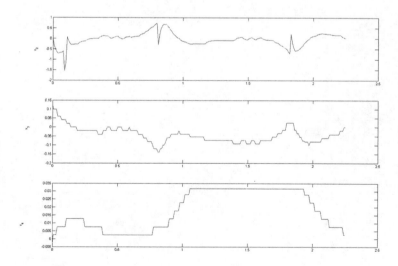

FIGURE 11.11 The car's states resulting from the using the ϕ estimator.

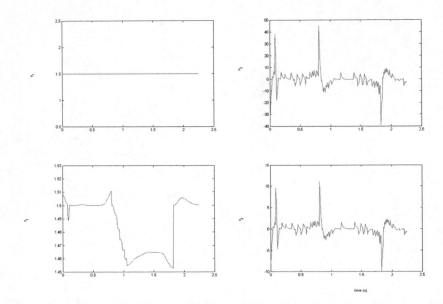

FIGURE 11.12 The control resulting from the using the ϕ estimator.

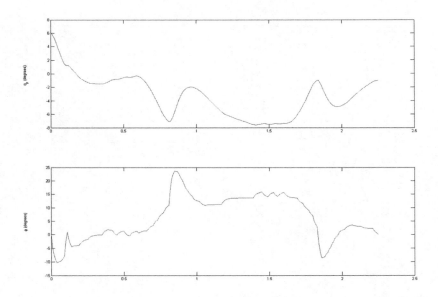

FIGURE 11.13 The heading angle θ_p and steering angle ϕ resulting from using the ϕ estimator.

In Fig. 11.12, spikes are seen where the path's curvature does not transition between 0 and $1/R$. As seen before, the ϕ estimator produced a false curvature while it was on the straightaway, due to transients while the car corrected itself. This can also be seen in the controller in Fig. 11.12. The two spikes occur in u_1 as it is starting out because the curvature is incorrectly estimated to be $1/R$ The first spike occurs at the 0 to $1/R$ transition, while the second occurs at the $1/R$ to 0 transition. After the initial transients, the controller performed similar to the previous controller with which the actual curvature was used.

Control using the model estimator

Next, the simulation was run using the model estimate method. The performance of this estimator, as shown in the previous chapter, was very good. As a result, there is very little difference between the controller's performance using the actual curvature or the estimated curvature. The results are shown in Figs. 11.14 through 11.06. This estimator was tested in both its forms. The form in which $\hat{a}$ is thresholded to obtain c is shown here. The form in which $\hat{a}$ itself was thresholded was also tested and produced identical results in the controller performance.

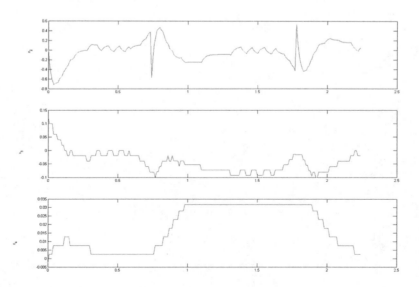

FIGURE 11.14 The car's states resulting from the using the model estimator.

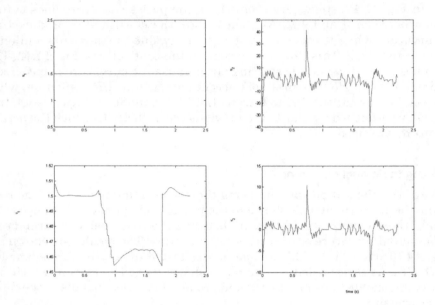

FIGURE 11.15   The control inputs resulting from the using the model estimator.

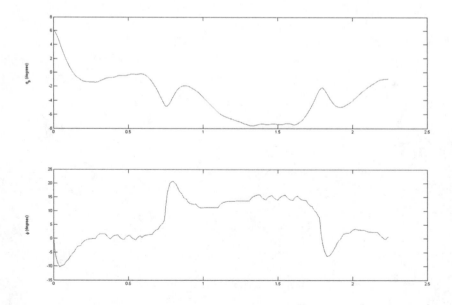

FIGURE 11.16   The heading angle θ_p and steering angle ϕ resulting from using the model estimator.

Appendix

A

Bibliography

Bay, John S. *Fundamentals of Linear State Space Systems*. Boston, MA: WCB/McGraw-Hill, 1999.

Bender, Edward A. *An Introduction to Mathematical Modeling*. New York: John Wiley & Sons, 1978.

Chapman, Stephen J. *Electric Machinery Fundamentals,* Fourth Edition. New York: McGraw-Hill, 2003.

De Luca, A., G. Oriolo, and C. Samson. "Feedback Control of a Nonholonomic Car-like Robot." In *Robot Motion Planning and Control,* edited by Jean-Paul Laumond. New York: Springer, 1998. Also available online at http://www.laas.fr/~jpl/book.html.

Desantis, R. M. "Path-Tracking for Car-Like Robots with Single and Double Steering." *IEEE Tran. Veh. Technol.* 4, No. 2 (May 1995).

Hayt, William H., Jr., and John A. Buch. *Engineering Electronmagnetics*, Sixth Edition. New York: McGraw-Hill, 2001.

James, D. J. G., and J. J. McDonald. *Case Studies in Mathematical Modelling*. New York: John Wiley & Sons, 1981.

Kachroo, P. "Microprocessor-Controlled Small-Scale Vehicles for Experiments in Automated Highway Systems." *The Korean Transport Policy Review* 4, No. 3 (1997): 145–178.

Kachroo, P. *Nonlinear Control Strategies and Vehicle Traction Control*. Ph.D. Dissertation. University of California at Berkeley, 1993. www.ee.vt.edu/~pushkin/PushkinKachroo_Berkeley_PhD.pdf

Kachroo, P., and M. Tomizuka. "Design and Analysis of Combined Longitudinal Traction and Lateral Vehicle Control for Automated Highway Systems Showing the Superiority of Traction Control in Providing Stability During

Lateral Maneuvers." *IEEE International Conference on Systems, Man, and Cybernetics*. 1995.

Khalil, Hassan K. *Nonlinear Systems*. Upper Saddle River, NJ: Prentice Hall, 1996.

Krause, P. C., and O. Wasynczuk. *Electromechanical Motion Devices*. New York: McGraw-Hill, 1989.

Kreyszig, Erwin. *Advanced Engineering Mathematics*. New York: John Wiley & Sons, 1993.

Lambert, A., T. Hamel, and N. Le Fort-Piat. "A Safe and Robust Path Following Planner for Wheeled Robots." *Proc. Intl. Conf. on Intelligent Robots and Systems*.Victoria, BC: IEEE/RSJ, October 1998.

Ma, Y., J. Kosecka, and S. Sastry. "Vision Guided Navigation for a Nonholonomic Mobile Robot." *IEEE Trans. on Robotics and Automation* 13, No. 3 (June 1999): 521–536

Martin, F. G. *Robotic Explorations*. Upper Saddle River, NJ: Prentice Hall, 2001.

Microchip Technology, Inc. *PIC16F87X Data Sheet*. Literature Number: DS30292C. 2001.

National Semiconductor, *A DMOS 3A, 55V, H-Bridge: The LMD18200*. Application Note 694. Tim Regan, author. December 1999.

Phillips, Charles L., and Royce D. Harbor. *Feedback Control Systems*. Upper Saddle River, NJ: Prentice Hall, 1988.

Pratap, Rudra. *Getting Started with Matlab: Version 6: A Quick Introduction for Scientists and Engineers*. Oxford: Oxford University Press, 2001.

Predko, Myke. *Programming and Customizing the PIC Microcontroller*. New York: McGraw-Hill, 1998.

Redfern, D., and C. Campbell. *The MATLAB 5 Handbook*. New York: Spriner-Verlag, 1998.

Reed, T. B. "Discussing Potential Improvements in Road Safety: A Comparison of Conditions in Japan and the United States to Guide Implementations of Intelligent Road Transportation Systems." *IVHS Issues and Technology,* SP-928 (1992): 1–12.

Rizzoni, G. *Principles and Applications of Electrical Engineering,* Third Edition. New York: McGraw-Hill, 2000.

Rohrs, Charles E., James L. Melsa, and Donald G. Schultz. *Linear Control Systems*. New York: McGraw-Hill, 1993.

Rudin, Walter. *Principles of Mathematical Analysis*. New York: McGraw-Hill, 1976.

Samson, C. "Control of chained systems. Application to path following and time-varying point-stabilization of mobile robots." *IEEE Trans. on Automatic Control* 40, No. 1 (1995): 64–77.

Shier, D. R., and K. T. Wallenius. *Applied Mathematical Modeling*. Boca Raton, FL: Chapman & Hall/CRC, 2000.

Slotine, J., and Weiping Li. *Applied Nonlinear Control*. Englewood Cliffs, NJ: Prentice Hall, 1991.

Texas Instruments. *TPIC0108B: PWM Control Intelligent H-Bridge*. Literature Number SLIS068A. April 2002.

"Traffic Safety Facts 2000: A Compilation of Motor Vehicle Crash Data from the Fatality Analysis Reporting System and the General Estimates System." DOT HS 809 337, U.S. Department of Transportation, National Highway Traffic Safety Administration, National Center for Statistics and Analysis. Washington, D.C., December 2001.

Ulaby, Fawwaz T. *Fundamentals of Applied Electromagnetics*. Upper Saddle, NJ: Prentice Hall, 2004.

Utter, D. "Passenger Vehicle Driver Cell Phone Use Results from the Fall 2000 National Occupant Protection Use Survey." Research Note, DOT HS 809 293, U.S. Department of Transportation, National Highway Traffic Safety Administration, July 2001.

Young, H. D., and R. A. Freedman. *Sears and Zamansky's University Physics, Tenth Edition*. San Francisco: Addison Wesley, 2000.

B

Schematics

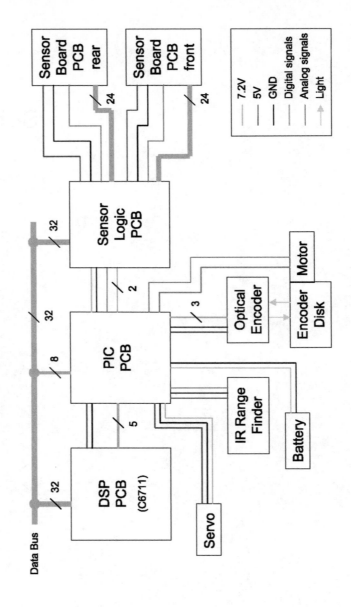

FLASH CAR INTERCONNECTIONS

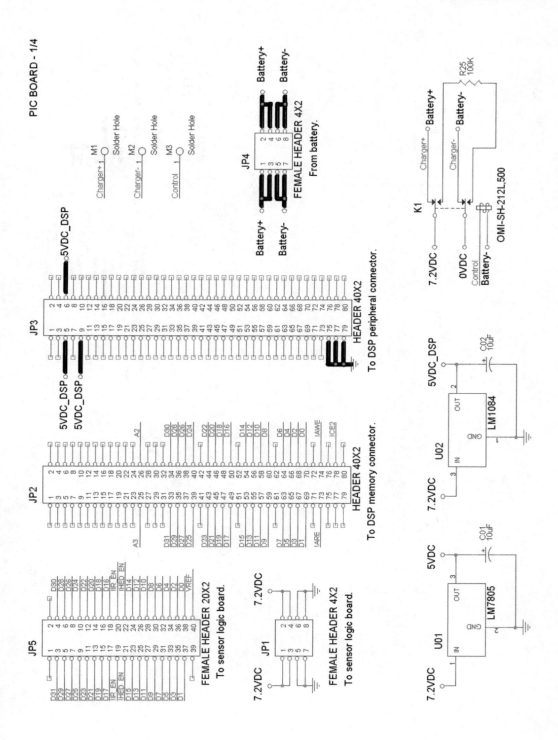

PIC BOARD - 1/4

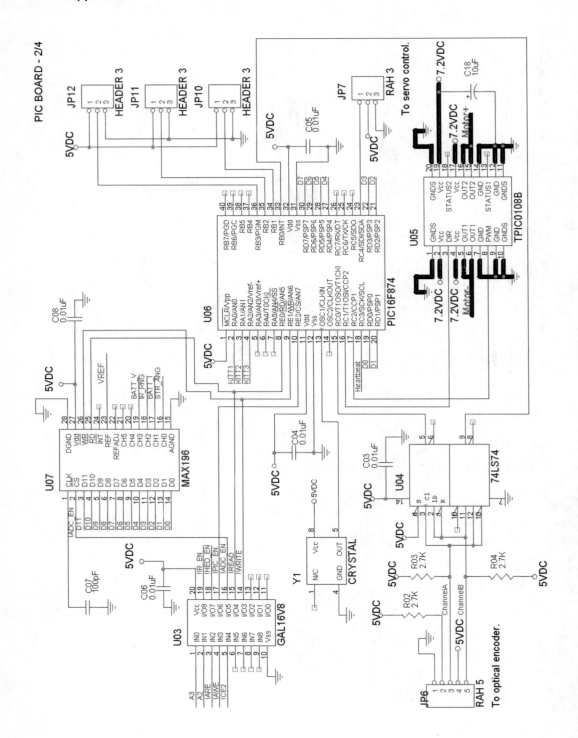

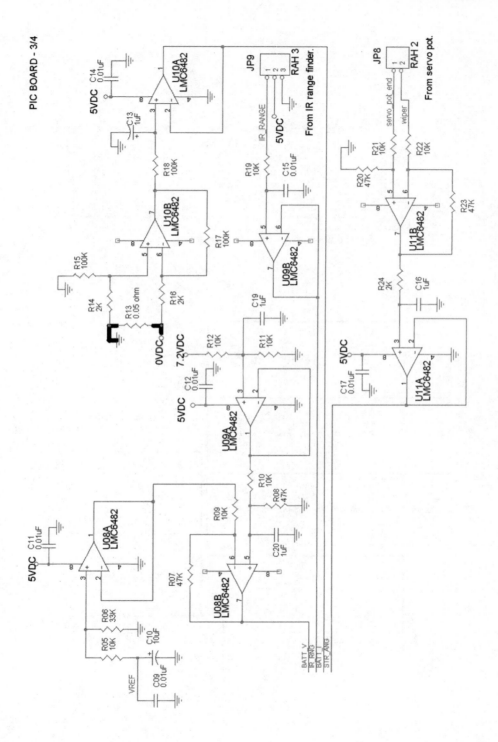

PIC BOARD - 3/4

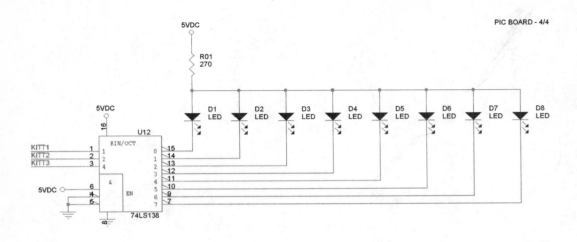

PIC BOARD - 4/4

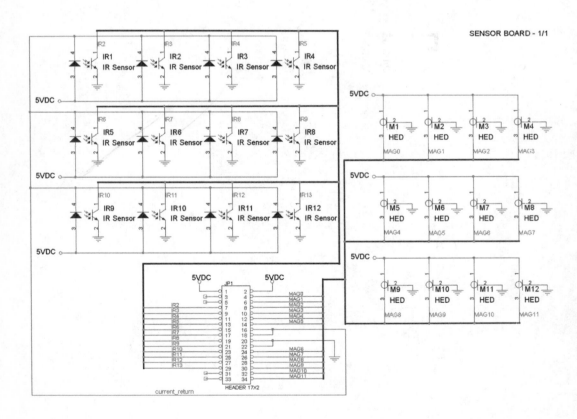

SENSOR BOARD - 1/1

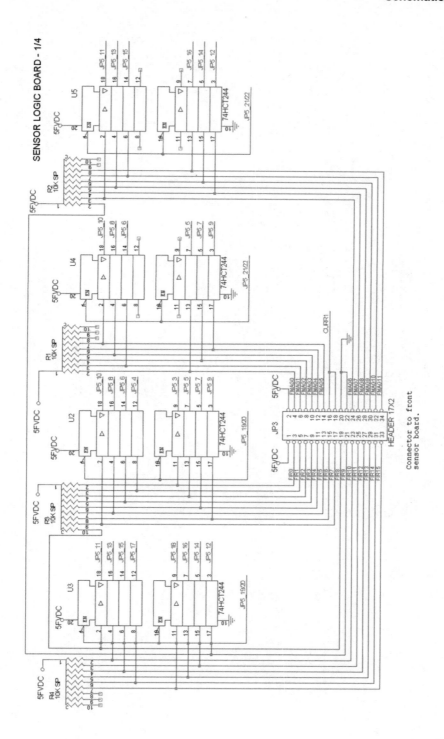

SENSOR LOGIC BOARD - 1/4

Connector to front sensor board.

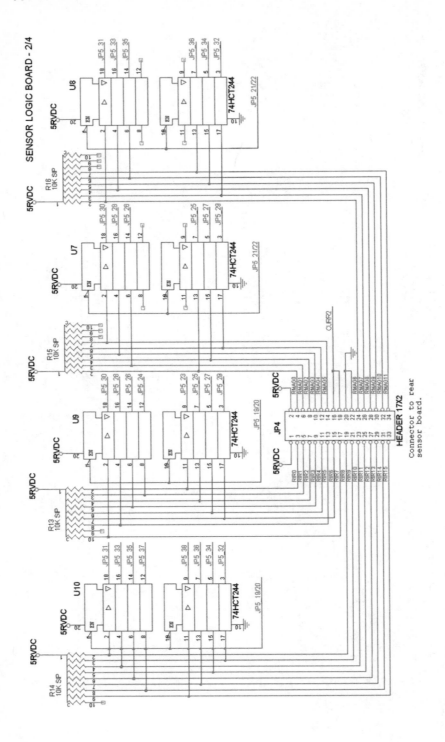

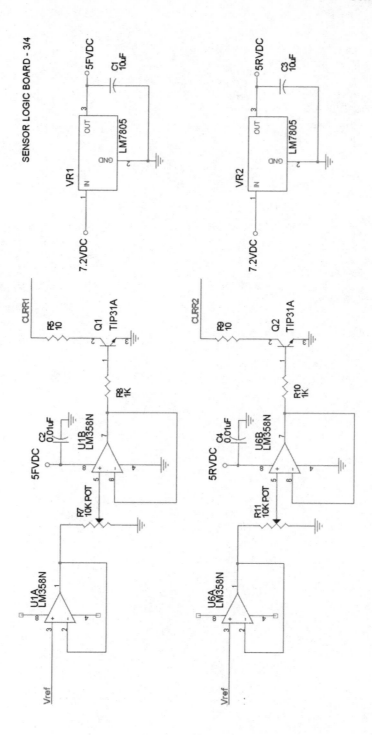

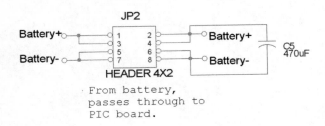

From battery,
passes through to
PIC board.

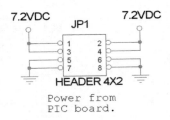

Power from
PIC board.

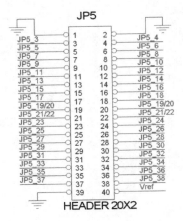

Parts Information

Flash Car Parts

Parts	Manufacturer	Manufacturer Part #	Vendor	Vendor Part #	Quantity
7.2V NiMH battery			Radio Shack	23-431	1
C6711 DSP starter kit	Texas Instruments	TMDS320006711			1
Optical disk, 1"	US Digital	I00	US Digital		1
Optical encoder module	US Digital	HEDS-9140-I00	US Digital		1
Trinity P2k Pro Stock motor*	Trinity	RC2117	Tower Hobbies	LXRR49	1
RC Legends Coupe kit	Bolink	BL 1342C	Tower Hobbies	LX8715	1
Standard servo	Futaba	S3003	Tower Hobbies	FUTM0031	1
IR range finder	Sharp	GP2D12	Acroname	R48-IR12	1
Battery connector set			Radio Shack	23-444	2
Power switch			Jameco	202948	1
Pinion Gear 48P 14T	Robinson Racing	1014	Tower Hobbies	LXEX12	1
Motor connector set	Duratrax	DTXC8263	Tower Hobbies	LXAMH2	2
Inline fuse holder			Jameco	102867	1
10-amp fast-acting fuse			Jameco	69462	1
CPU fan	Sunon	KDE0502PFB3-8	Jameco	206156	1
PIC board					1
Sensor board					2
Sensor logic board					1

*motor rewound with 100 turns of 30 AWG wire

Pic Board Parts

Parts	Manufacturer	Manufacturer Part #	Vendor	Vendor Part #	Quantity
10uF electrolytic capacitor		R10/50	Jameco	29891	4
0.01uF ceramic capacitor		MD.01	Jameco	25507	11
100pF ceramic capacitor		MD100	Jameco	81525	1
1uF ceramic capacitor		MD1	Jameco	81509	4
Red diffused LED		LA93B/H-1	Jameco	104248	8
2x4x2 female 0.1" header			Jameco	70826	1
40x2 female SMT header	Samtec	TFM-140-32-S-D-A	Samtec		2
20x2 female 0.1" header			Jameco	111704	1
36-pin right angle 0.1" header					
5-pin connector housing		SCH5	Jameco	163686	2
3-pin connector housing		SCH3	Jameco	157382	1
2-pin connector housing		SCH2	Jameco	100811	1
Female pins		FCH1	Jameco	100765	15
3-pin male 0.1" header		SMH03	Jameco	109575	3
DPDT relay	Tyco OEG	OMI-SH-205L,594			1
270 ohm resistor			Jameco	30605	1
2.7k ohm resistor			Jameco	30390	3
10k ohm resistor			Jameco	29911	8
33k ohm resistor			Jameco	30841	1
47k ohm resistor			Jameco	31149	4
0.05 ohm resistor	Ohmite	12FR050	Digikey	12FR050-ND	1
2k ohm resistor			Jameco	30277	3
100k ohm resistor			Jameco	29997	3
5V SMT regulator	National Semiconductor	LM340S-5.0	Digikey	LM340S-5.0-ND	1
5V 5A SMT regulator	National Semiconductor	LM1084IS-5.0	Digikey	LM1084IS-5.0-ND	1
CMOS PLD	Lattice	GAL16V8-25QNC	Jameco	37461	1
Dual D flip-flop	Fairchild Semiconductor	DM74LS74AN	Jameco	48004	1
3 to 8 decoder	Fairchild Semiconductor	DM74LS138N	Jameco	46607	1
PWM control H-bridge	Texas Instruments	TPIC0108B	Digikey	96-10857-5-ND	1
H-bridge heatsink	Aavid Thermalloy	580200B00000	Digikey	HS179-ND	1
20-pin SOIC-DIP converter	Aries Electronics	20-350000-10	Digikey	A727-ND	1
CMOS PIC microcontroller	Microchip	PIC16F874-20/P	Digikey	PIC16F874-20/P-ND	1
12-bit A/D converter	Maxim	MAX196ACNI	Digikey	MAX196ACNI-ND	1
Dual Rail-to-Rail Opamp	National Semiconductor	LMC6482IN	Digikey	LMC6482IN-ND	4
20 MHz crystal oscillator	ECS Inc.	ECS-2200B-200	Digikey	XC274-ND	1
20-pin DIP socket		20MLP	Jameco	38623	1
40-pin DIP socket		40MLP	Jameco	41136	1

Sensor Logic Parts

Parts	Manufacturer	Manufacturer Part #	Vendor	Vendor Part #	Quantity
10uF electrolytic capacitor		R10/50	Jameco	29891	2
0.01uF ceramic capacitor		MD.01	Jameco	25507	5
4x2 male 0.1" stackable header	Samtec	DW-04-12-G-D-655			2
17x2 male 0.1" header		923864R	Jameco	53516	2
20x2 male 0.1" stackable header	Samtec	DW-20-12-G-D-655			1
NPN transistor	Fairchild Semiconductor	TIP31A	Jameco	33048	2
10k ohm SIP resistor			Jameco	24643	8
1k ohm resistor			Jameco	29663	4
10 ohm resistor			Jameco	29882	2
10k ohm potentiometer	Bourns	3266W-LTC-103	Digikey	3266W-103-ND	2
Low power dual opamp	National Semiconductor	LM358N	Jameco	23966	2
Octal tri-state buffer	Texas Instruments	SN74HC244N	Jameco	45022	8
5V SMT regulator	National Semiconductor	LM340S-5.0	Digikey	LM340S-5.0-ND	2
8-pin DIP socket		8MLP	Jameco	51625	2
20-pin DIP socket		20MLP	Jameco	38623	8
34-pin ribbon connector	CW Industries	CWR-220-34-0000	Digikey	CSC34G-ND	2

Sensor Board Parts

Parts	Manufacturer	Manufacturer Part #	Vendor	Vendor Part #	Quantity
IR reflective object sensor	Fairchild Semiconductor	QRD1114	Digikey	QRD1114-ND	12
17x2 male 0.1" header		923864R	Jameco	53516	1
Hall effect sensor	Micronas	HAL506UA-E	Symmetry Electronics Corporation		12
34-pin ribbon connector	CW Industries	CWR-220-34-0000	Digikey	CSC34G-ND	1
34-conductor ribbon cable			Jameco	37840	1

Hardware Sources

Company:	Aavid Thermalloy
Product:	Heat sinks
Address:	80 Commercial Street
	Concord, NH 03301
Telephone:	1-603-224-9988
Internet:	www.aavidthermalloy.com

Company:	Acroname, Inc.
Product:	Robotic component supplier
Address:	4894 Sterling Drive
	Boulder, CO 80301-2350
Telephone:	1-720-564-0373
Internet:	www.acroname.com

Company:	Advanced Circuits
Product:	Printed circuit boards
Address:	21101 East 32nd Parkway
	Aurora, CO 80011
Telephone:	1-800-979-4PCB
Internet:	www.4pcb.com

Company:	Aries Electronics, Inc.
Product:	IC adapters
Address:	P.O. Box 130
	Trenton Avenue
	Frenchtown, NJ 08825
Telephone:	1-908-996-6841
Internet:	www.arieselec.com

Company: Bolink
Product: RC car chassis
Address: 420 Hosea Road
             Lawrenceville, GA 30245
Telephone: 1-770-963-0252
Internet: www.bolink.com

Company: Bourns, Inc.
Product: Potentiometers
Address: 1200 Columbia Avenue
             Riverside, CA 92507-2114
Telephone: 1-877-4-BOURNS
Internet: www.bourns.com

Company: Digi-Key Corporation
Product: Electronic components
Address: 701 Brooks Avenue South
             Thief River Falls, MN 56701
Telephone: 1-800-DIGI-KEY
Internet: www.digikey.com

Company: ECS, Inc. International
Product: Crystal oscillators
Address: 1105 South Ridgeview
             Olathe, KS 66062
Telephone: 1-800-237-1041
Internet: www.ecsxtal.com

Company: Fairchild Semiconductor Corporation
Product: Infrared sensors
Address: 82 Running Hill Road
             South Portland, ME 04106
Telephone: 1-800-341-0392
Internet: www.fairchildsemi.com

Company: Futaba Corporation of America
Product: Servos
Address: 2865 Wall Triana Highway
             Huntsville, AL 35824
Telephone: 1-256-461-7348
Internet: www.futaba.com

Company: Jameco Electronics
Product: Electronic components
Address: 1355 Shoreway Road
             Belmont, CA 94002-4100
Telephone: 1-800-831-4242
Internet: www.jameco.com

Company: Lattice Semiconductor Corporation
Product: Programmable logic devices
Address: 5555 N.E. Moore Court
            Hillsboro, OR 97124
Telephone: 1-503-268-8000
Internet: www.latticesemi.com

Company: Maxim Integrated Products, Inc.
Product: Analog-to-digital converters
Address: 120 San Gabriel Drive
            Sunnyvale, CA 94086
Telephone: 1-800-998-9872
Internet: www.maxim-ic.com

Company: Microchip Technology, Inc.
Product: PIC microcontrollers
Address: 2355 West Chandler Boulevard
            Chandler, AZ 85224-6199
Telephone: 1-480-792-7200
Internet: www.microchip.com

Company: Micronas Semiconductor Holding AG
Product: Hall effect sensors
Address: Technopark
            Technoparkstrasse 1
            CH-8005 Zurich
            Switzerland
Telephone: +41-1-445-3960
Internet: www.micronas.com

Company: National Semiconductor Corporation
Product: Discrete semiconductor components
Address: 2900 Semiconductor Drive
            P.O. Box 58090
            Santa Clara, CA 95052-8090
Telephone: 1-408-721-5000
Internet: www.national.com

Company: Ohmite Mfg. Co.
Product: Current sense resistors
Address: 1600 Golf Road, Suite 850
            Rolling Meadows, IL 60008
Telephone: 1-866-9-OHMITE
Internet: www.ohmite.com

Company: RadioShack Corporation
Product: RC car accessories
Address: 300 West Third Street, Suite 1400
              Fort Worth, TX 76102
Telephone: 1-800-THE SHACK
Internet: www.radioshack.com

Company: Samtec, Inc.
Product: Connectors
Address: 520 Park East Boulevard
              New Albany, IN 47150
Telephone: 1-800-SAMTEC-9
Internet: www.samtec.com

Company: Sharp Microelectronics
Product: Infrared range finder
Address: 5700 NW Pacific Rim Boulevard
              Camas, WA 98607
Telephone: 1-360-834-2500
Internet: www.sharpusa.com

Company: Symmetry Electronics Corporation
Product: Hall effect sensors
Address: 5400 Rosecrans Avenue
              Hawthorne, CA 90250
Telephone: 1-310-536-6190
Internet: www.symmetryla.com

Company: Texas Instruments Incorporated
Product: Digital signal processors and discrete logic
Address: 12500 TI Boulevard
              Dallas, TX 75243-4136
Telephone: 1-800-336-5236
Internet: www.ti.com

Company: Tower Hobbies
Product: RC car components
Address: PO Box 9078
              Champaign, IL 61826-9078
Telephone: 1-800-637-6050
Internet: www.towerhobbies.com

Company: Trinity Products, Inc.
Product: RC electric motors
Address: 36 Meridian Road
              Edison, NJ 08820
Telephone: 1-732-635-1600
Internet: www.teamtrinity.com

Company: Tyco Electronics
Product: Relays
Address: P.O. Box 3608
              Harrisonburg, PA 17105
Telephone: 1-717-564-0100
Internet: www.tycoelectronics.com

Company: U.S. Digital Corporation
Product: Optical encoders
Address: 11100 N.E. 34th Circle
              Vancouver, WA 98682
Telephone: 1-800-736-0194
Internet: www.usdigital.com

Index